50.–

III only

BEIHEFT 102 ZUR NOVA HEDWIGIA

BEIHEFTE ZUR

NOVA HEDWIGIA

HEFT 102

The Boletineae of Mexico and Central America III

by

R. SINGER, J. GARCIA & L.D. GOMEZ

with 23 figures and 1 plate

J. CRAMER

in der Gebrüder Borntraeger Verlagsbuchhandlung

BERLIN · STUTTGART 1991

Author's address:

Rolf Singer
Field Museum of Natural History
Chicago, Ill. 60605, U.S.A.

Jesus García
Instituto Tecnológico de Cd. Victoria
Apdo. Postal 175, Cd. Victoria, Tamaulipas, Mexico

Luis Diego Gomez
Jardín Botánico de Las Cruces
Apdo 35, San Vito de Coto Brus, Costa Rica

ISBN 3-443-51024-8

Printed in Germany by Proff Offsetdruck, 8196 Eurasburg

Contents

Summary 1

Key to the genera of the region
(Boletoideae with pink, chestnut or yellow spore print). 3

I. *Veloporpyrellus* 5

II. *Tylopilus* 8

III. *Austroboletus* 53

IV. *Porphyrellus* 60

V. *Fistulinella* 73

VI. *Xanthoconium* 85

Diagnoses latinae 88

Acknowledgments 91

Supplement to part I: *Paxillus bambusinus* 92

Literature cited in part III 94

Index to taxa 96

Plates 99

Summary

In this part of Boletineae-monographs for the area between California (U.S.A.) Colombia we have treated the subfamily Boletoideae of the family Boletaceae but only those genera with never olivaceous spore print. We are treating in *Veloporphyrellus* one species, in *Tylopilus* — 22 species, 11 of these new, in *Austroboletus* — 3 species, of these one new, in *Porphyrellus* 5 species, of these 1 new, in Fistulinella 5 species, of these two new, and in *Xanthoconium* one species. We did not attempt to divide *Tylopilus* into sections but lelieve that all species belong in section *Tylopilus* and sect. *Oxydabiles* Sing. and that the Brazilian (and some paleotropical) sections are different. In the region treated here they do not occur, yet *T. corneri* and a tropical Asiatic species without pseudocystidia may represent a new section. — *Porphyrellus* is a northern genus that reaches Mexico while *Tylopilus* is a genus of tropical origin that reaches the temperate zone but with fewer species as is moves northwards. The only genus (of those treated in this part) that seems to be restricted to Mexico and Central America is *Veloporphyrellus*.

Key to the genera of the region
(Boletoideae with non-olivaceous spore print)

1. Stipe scabrous, i.e. the hymeniform fascicles of the stipe not directly formed on the parallel or subparallel hyphae of the fundamental hyphae of the stipe cortex, but as end-cells of strands of hyphae diverging from the stipe tissue. LECCINUM (not treated here).

1. Stipe not scabrous but pustulate to pruinate, tomentose, or glabrous, often partially of entirely reticulated, or smooth to subfibrillose.

    2. Pseudocystidia in the hymenophore present, or at least present as dermatocystidia.

        3. Spores, at least in SEM with distinct ornamentation (see "8" below).

        3. Spores smooth.

            4. Pileus covered by an ixotrichodermium; pileus pink (see "6" below).

            4. Pileus covered by a trichodermium, a trichodermial palisade, or an epithelium, the trichodermium sometimes eventually partly cutis-like.

                5. Context and/or hymenophore generally bluing when quite fresh and young; spores often longer than 15.5 μm; ectomycorrhizal with conifers (Pinaceae) or Fagales, perhaps *Leptospermum*. Not observed in Central America.......... IV. PORPHYRELLUS p. 60

                5. Context and hymenophore never bluing when injured; spores rarely (and then only as relatively rare giant spores) longer than 15.5 μm; ectomycorrhizal mostly with *Quercus*. Common in Central America. .... II. TYLOPILUS p. 8

                (Pseudocystidia soon becoming coscinoid; spores < 8 μm long, $Q < 2$; slow-growing carpophores with bitter taste. Tropical Asia. BOLETOCHAETE).

    2. Hymenophore devoid of pseudocystidia and dermatocystidia not pseudoamyloid.

        6. Pileus covered with an ixotrichodermium, at least when fresh and quite mature or with a gelatinascent epicutis area, then mostly with scrobiculate pileus (cf. "9" if spore print ferruginous)..... V. FISTULINELLA p. 93

        6. Pileus without an ixotrichodermium, quite dry, epicutis not gelatinizing.

            7. Veil present; carpophore pigmentless (except spores)................ .... I. VELOPORPHYRELLUS p. 5

7. Veil absent; carpophore pigmented.

   8. Spores ornamented. . . . . III. AUSTROBOLETUS p. 53

   8. Spores not or not distinctly ornamented (SEM).

      9. Spore print yellow; pileus dry; spores narrow. . . . . VI. XANTHOCONIUM p. 85

      9. Spore print pink to chestnut (see TYLOPILUS, supplement p. 49) or ferruginous and then pileus glutinous (PULVEROBOLETUS, not treated here).

I. VELOPORPYRELLUS Sing.

Brenesia, 22: 293. 1984

This is a monotypic genus, the characters evident from the key above.

Veloporphyrellus pantoleucus Sing., l.c. (Pl. 1, Fig. 1-4; pl. 2, fig. 5-8).

Pileus white or dirty pale grayish in center only, smooth, rarely very slightly faveolate-uneven, dry (not viscid), finely and thinly but distinctly tomentose, more rarely becoming subtomentose or with cinnamon-pallid linear fibrils all over, with at first acute or subacute, projecting margin which continues into the white, membranous, dry; smooth veil connecting it with the apex of the stipe, at first pulvinate-convex then conic-convex, obtuse or convex-umbonate, 34-52 mm broad.

Hymenophore tubular, tubes very pale pinkish (independently of the spore masses), then pink, later light pinkish-pearl-gray but not changing color when bruised, 5-11 mm long where longest (hymenophore somewhat ventricose), depressed around the stipe in mature specimens and tending to be narrowly sublamellar where attached to the stipe; pores concolorous, slightly dirtier when touched, small (up to 0.5 mm wide in dried material, not reaching 1 mm in diam. in fresh material, tubes not separating in alcohol preparations), pores mostly simple, few compound, some roundish, some angular, not stuffed when young. Spore print ocher-purple becoming more grayish by dehydration.

Stipe white or with some pale brownish hues at the base, pubescent at the apex, below apex under a lens distantly to indistinctly pubescent, smooth and macroscopically neither scabrous nor pustulate or reticulate anywhere, solid, equal in upper third or all over, or gradually widening to the base which is then short attenuate downwards, 48-120 × 6-17 mm; veil at first attached to apex of the stipe and mostly somewhat recurved but no persistent annular veil left; basal mycelium white.

Context white, unchanging but sometimes staining a very pale burgundy red in the periphery of the pileus, firm, especially in the stipe; odor somewhat like bread; taste submild (leaving a faint bitter or acrid aftertaste).

Chemical characters: Sulfovanillin and phenol: Negative. — $FeCl_3$ slowly lightly dirty green, — NH_4OH negative but sometimes accelerating the reddening of the pileus context.

Spores 7.5-13.5 × 3-4.5 μm, mostly 11.3-13.2 × 3.3-4.2 μm (× = 12.31 × 4.10 μm), mostly fusoid, rarely ellipsoid, oblong or sausage shaped, mostly with a suprahilar depression, with moderately thick wall, all perfectly smooth under the light microscope, pale melleous to subhyaline, many distinctly pseudoamyloid, but some weakly or not pseudoamyloid, cyanophilous.

Hymenium: Basidia (19)-28-32 × 6-9.5 μm, in young specimens 2-3-4-spored, soon all 4-spored. Cystidia very numerous at the pore level (and there visible under a hand lens), less numerous but not sparse inside the tubes, versiform, most frequently ampullaceous, 40-75 × 4.5-10 μm with the apex 10-20 × 3.5-4 μm obtuse or else fusoid or subfusoid, utriform, or clavate, (15)-40-70 × (3.5)-4.5-11 μm acute, subacute or obtuse, thin-walled, not incrusted, wall and interior inamyloid, hyaline in KOH, under the light microscope appearing "empty" (optically) or very finely granular to guttulate inside the cheilocystidia, rarely or frequently internally 1-2 septate.

Hyphae: Hymenophoral trama bilateral of the *Boletus*-type, hyaline, the hyphae of the mediostratum subinterwoven to subparallel and non-gelatinized, 3-4.5 μm broad, those of the lateral stratum well diverging, distinctly gelatinized and not touching each other, after reaching maturity denser and less recurved: all hyphae inamyloid, without clamp connections.

Cortical layers: Epicutis of the pileus trichodermial, non-gelatinized, consisting of long septate (primary and secondary septa) hyphae and hyphal hairs (especially near margin of pileus) such as found on the surface of the stipe, all these elements without pigment or incrustation, inamyloid. Covering of the stipe consisting of hair-like hyphal extensions causing the pubescence of the apex of the stipe and patches or bunches of subhymeniformly arranged dermatobasidia, dermatocystidia, and other dermatocystidioid elements; dermatobasidia and dermatobasidioles rather numerous near apex, scattered further below on the stipe, 17-38 × 6.6 - 8 μm, the dermatobasidia 2-3-4-spored producing spores 7.5-11 × 3.7-4.5 μm, ellipsoid, oblong or fusoid, otherwise like those produced by the hymenophore; dermatocystidia of three types, (I) 12.5-20 × 5.5-8.5 μm, basidiomorphous, mostly clavate but some more ventricose or fusoid, all with rounded-obtuse tip, very numerous both at apex and lower portion of stipe, many of them with an applicate or broken-off incrustation, others non-incrusted and thin-walled, interior, wall, and incrustation hyaline or subhyaline; (II) 28-43 × 5.5-8.5-(10) μm, ampullaceous, entirely hyaline without visible contents, thin-walled, at apex of stipe also longer (reaching 70 × 11 μm) and sometimes utriform, fusoid, or subcylindrical; (III) 30-57 × 8.5-10 μm, clavate, broadly fusoid, or ampullaceous, mostly firm-walled, with gelatinizing wall (the gelatinizing wall surfaces and wall-incrustations of dermotocystidia type III believed to be responsible for the subviscid "feel" of the stipe surface much like the glands of some Suilli). In and between the gland-like patches the piliform hyphae reach 125 μm, end cell often cystidiform.

Chemical color reactions: KOH on pileus yellowish brown, slightly pale green, on hymenophore and stipe slightly pale yellow — NH_4OH and $FeSO_4$ negative.

On the ground in *Quercus-Magnolia* forest of the tropical-montane zone, at about 1800-2000 m altitude in Costa Rica, 1200-1800 m in Mexico.

Material studied: COSTA RICA: Cartago, 1.5 km SE of junction of dirt road to San Cristóbal (Casamata) and route 2, XI 1983, L.G. Gómez 22007 (F), typus with photo, dried and alcohol material. — San Cristóbal VII 1983, Gómez & Alfaro 21232 (F). — Same place as type, 30 VII 1986, Singer B 14529 (F), topotype. — MEXICO: Mexico, Mun. Tejupilco, Nanchititla, 2 VII 1988 González-Velásquez 736, det. García, (F, ENCB) — Hidalgo, km 168 road Pachuca-Tampico 16-VII 1988, A.D. Guevara 178, det. García (F, FCME).

II. TYLOPILUS Karst.

Rev. Mycol. 3: 16. 1881.

Syn: *Rhodoporus* (Quél.) Bat., Bolets p. 11. 1908
Rhodobolites G. Beck, Zeitschr. f. Pilzk. 2: 146. 1923.
Leucogyroporus Snell, Mycologia 34: 408. 1942.

The genus as here monographed for the region of Meso-America and Central America is taken with the delimitation suggested by Wolfe (1979 as subgenus *Tylopilus* and Singer (1986), i.e., with true pseudocystidia or at least with pseudoamyloid incrusting granules, with pink to chestnut colored spore print and constantly ectomycorrhizal; the pores are generally small and the hymenophoral trama is os the *Boletus*-type. Species with veil and/or without pseudocystidia, and species with (in youth) hymenophoral trama of the Phylloporus-type are not admitted. Those with scabrous stipe are transferred to *Leccinum;* those with an ixotrichodermium are transferred to *Fistulinella.* Our monograph shows that the South American species and sections do not occur further north, and the two sections *Tylopilus* and *Ferruginei* Sing. should be redefined since they do not seem to show a clear natural hiatus between them. The following key is for a combination of species belonging to these two sections.

It appears that two pathways have been followed by *Tylopilus* species, one through tropical Asia (and Africa?), and one through the Americas, with two centers of distribution (tropical Asia and Central America). This distribution is quite different from that of *Porphyrellus,* a typically north-temperate, occasionally subtropical, genus which has not been observed south of Mexico.

Key to the species of the region

1. Pileus orange; spores, at least many of them remarkably short and broad. Mexico. . . . . 1. T. BALLOUI

1. Pileus never orange; spores, even if relatively small, not very broad.

2. Pileus viscid in age (see genus *Fistulinella*).
2. Pileus dry; epicutis not gelatinascent.

3. Epicutis in young and fresh material consisting of a trichodermial palisade or stipe with a hymeniform covering containing spherocystoid cells.

4. Taste bitter; mycorrhiza with *Quercus* (if with *Pinus,* see "13").

5. Surfaces when handled and context where bruised not tending to become black or blackish.

6. Pileus violet; stipe also mostly violet; epicutis distinctly palisadic, consisting of dermatocystidia rising from a trichodermium, the dermatocystidia ampullaceous, lanceolate, more rarely cylindrical, 12-44 μm long with obtuse to subacute tip. Mexico. . . . . 2. T. PLUMBEOVIOLACEUS

6. Color different; epicutis not strictly palisadic and dermatocystidia not developed, or different.

7. Context unchanging (see "19").

7. Context not unchanging.

8. Elements of the covering layer of the stipe consisting of spherocystoid elements, these often in chains and mostly 9-13 μm broad. Spores 6.3-9.5(10) μm long. . . . . 3. T. JAPALENSIS

8. Elements of stipe covering different (see "17").

5. Surfaces when handled and context when bruised at least partially blackening (see "10").

4. Taste mild.

9. Stipe at first white, then ± cream colored; basal mycelium white, apex of stipe hardly more than 15 mm wide; spores 9-13 × 3.5 μm; contents of epicutis-cells very weakly or not pseudoamyloid. . . . . 4. T. MITISSIMUS

9. Stipe concolorous with pileus or in lower half soon becoming so.

10. Pileus pinkish brown. Mexico. . . . . 5. T. MONTOYAE

10. Pileus darker.

11. Surfaces when handled and context distinctly changing (over pinkish gray) to black or blackish; spores to 13 μm long. North America to Mexico. . . . . . . 6. T. ALBOATER

11. Not blackening; spores reaching 14.8 μm. Costa Rica to Colombia. . . . . 7. T. OBSCURUS

3. Epicutis a trichodermium, not palisadic.

12. Ectomycorrhiza with *Pinus*.

13. Context not pink when bruised; taste either slightly or strongly bitter.

14. Spores 8-10.7 × 3-3.7 μm....... 8. T. NICARAGUENSIS

14. Spores 12-13.5(14.5) × 4-5(5.3) μm...20. HONDURENSIS

13. Context pink when bruised; spores 11-14.5 × 3.5-5.5(6) μm; taste unknown (see "24").

12. Ectomycorrhiza with *Quercus*.

15. Ectomycorrhiza with *Quercus oleoides* in lowland forests of Costa Rica; apex of stipe finely reticulated; spores 9-13 × 3.5 μm.

16. Strands of thin-fibrillose, fuliginous hyphae forming a thin loose felt above the livid-grayish ground of the pileus surface, context white, white turning to pink when bruised, then blackening; stipe ± concolorous with the pileus................ .... 9. SANCTAEROSAE

16. Pileus dark burnt umber or brown to yellowish brown; context white, unchanging; stipe raw umber, sienna or coffee brown or livid-maroon....... 10. T. GUANACASTENSIS

15. Mycorrhiza with other species of oak in the montane to subparamo zone.

17. Stipe considerably longer than the diameter of the pileus, obclavate, not reticulate or reticulate only at the very apex of the stipe; taste mild if context unchanging, or else with bitterish after-taste.

18. Context and hymenophore practically unchanging when bruised; pileus livid, later grayish brown; taste mild..... .... 11. T. LIVIDOBRUNNEUS

18. Context turning pink, or ochrascent; pileus umber or light umber on mostly chamois ground; taste with a bitterish after-taste................... 12. T. GOMEZII

17. Stipe versiform, mostly about as long as the diameter of the pileus, or shorter or slightly longer; taste bitter only if the context is unchanging.

19. Context unchanging; taste bitter; Mexican species.

20. Spores mostly 10.5-11.5 × ± 3.5 μm with some spores reaching 12.5 μm long, and usually a few giant spores up to 18 × 3.5-5 μm present; epicutis in the uppermost as well as in the lower layers composed of two kinds of cells, the elongated to subfilamentose

ones and the swollen ones, the latter often subiso-diametric and 10-17 μm diam.; stipe subbulbous below. . . . . 13. T. SUBCELLULOSUS

20. Spores mostly 8-9 × 4 μm, giant spores not observed; epicutis of pileus with elongated elements only, these not broader than 6 μm; stipe typically subequal, or ventricose in the middle. . . . . . 14. T. WILLIAMSII

19. Not at the same time with bitter as well as unchanging context.

21. Pileus either with purplish or violet tinge, or sepia to almost black.

22. Stipe either only reticulate in a few millimeters of the very apex, or in the upper third but then consisting of small pustules arranged in a reticulate pattern, in this zone context often colored; stipe obclavate, often basically appendiculate; context unchanging, or turning brown when injured. . . . . 15. T. COSTARICENSIS

22. Stipe distinctly reticulated, mostly down over the upper two thirds of the ventricose (at least in the lower half) stipe; context pink or pinkish gray when bruised, at least above hymenophore or in the apex of the stipe.

23. Pileus at first almost black; spores mostly 11-14 × 4-5.5 μm. . . . . 16. T. SUBNIGER

23. Pileus in youth not blackish; spores smaller. . . . . 17. T. VINACEOGRISEUS

21. Pileus neither with purple or violet tones, nor almost black when young.

24. Stipe reticulate either at the extreme apex only or down to one third or two thirds of the stipe length; spores 9.5-12.5 × 3-4(4.8) μm; cells of the epicutis up to 9, often to 13 μm diam. Widely distributed species from the east coast of North America to Florida and the Gulf coast, Mexico, and Japan. . . . . 18. T. FERRUGINEUS

24. Stipe mostly reticulated at the apex; below reticulation sometimes dotted; cells of the epicutis narrower; spores somewhat larger.

25. Stipe rather long and subcylindrical; reticulated at apex, below reticulated zone not

dotted with dried gray dots; dermatocystidia much like hymenial pseudocystidia. . . . . 19. T. INDECISUS

25. Stipe short and strongly tapering downwards; reticulated above, below reticulation dotted gray when dried; dermatocystidia of the stipe mostly clavate. . . . . 20. T. BRACHYPUS

1. **Tylopilus balloui** (Peck) Sing., Amer. Nidl.-Nat. 37: 232. 1947. (Pl.2, fig. 9-10; pl.3, 11-12).

   Boletus balloui Peck, Bull. N.Y. State Mus. 157: 22. 1912.
   Gyrodon balloui (Peck) Snell, Mycologia 33: 422. 1941.
   Boletus subsanguineus Peck sensu (Murr.) Coker & Beers, Bol. N. Car. p. 23. 1943.

Pileus orange (apricot orange to "bittersweet" or "flame scarlet" (Ridgway, according to Coker & Beers 1943), dried mostly rusty cinnamon, subglabrous, smooth or slightly pitted in the center, slightly viscid after rains, but generally dry, with slightly projecting margin, convex, but often becoming irregular or depresses in the center, 40-85 mm broad.

Hymenophore tubular, tubes whitish, only up to 9 mm long, somewhat depressed around the stipe or simply adnexed; pores concolorous with the tubes or cream to flesh color, or almost blood red, stained brown when touched, subisodiametric or near margin somewhat radially elongated, 0.5-1.5 μm diam. Spore print not obtained.

Stipe yellow or yellowish white apex, or all over, below often becoming brown, in the middle sometimes concolorous with the pileus, finely pruinose, glabrescent, dry, solid, characteristically tapering downwards, or cylindrical with widened apex and subacute base, smooth but often with the extreme apex lineate just below the hymenophore, but not reticulated, 25-75 × 8-18 mm, at the flaring to even reaching 30 mm across (Coker & Beers 1943); veil none; basal mycelium whitish.

Context white or whitish, often orangy in part of the stipe, in parts of the pileus turning ± dirty lilac to purple brown when young and fresh; taste not bitter; odor strong and unpleasant when dried.

Spores 6.5-11 × 3.5-5 μm, mostly 7-8.5 × 3.7-4.7 μm, (Q = 1.7, or ± = 2 (2.7), ellipsoid to ovoid or fusoid, frequently broadest below, often laterally bean-shaped, extremely variable, without, more rarely with a suprahilar depression, thin-walled, inamyloid or slowly very weakly pseudoamyloid in superposition, smooth without germ pore or truncation.

Hymenium: Basidia 24-27.5 (36) × 6.5-9 μm, (20) 4-spored. Pseudocystidia 27-60 (80) × 6.5-16.5 (21.5) μm, utriform and obtusely rounded at tip, with strongly pseudoamyloid granular to amorphous contents, but many versiform, fusoid and subacute or with subacute to acute appendage (1-11 μm long), with the contents weakly pseudoamyloid or inamyloid, weakly or sometimes not granular inside. True pleurocystidia and cheilocystidia not differentiated.

Hyphae inamyloid and without clamp connections. Hymenophoral trama bilateral of the *Boletus*-type, with a melleous mediostratum of non-gelatinized 2-4.5 μm thick hyphae, and the lateral stratum at first with rather strongly divergent and strongly gelatinized hyphae 2.7-8.3 μm thick. Oleiferous hyphae present.

Epicutis a trichodermium of elongated cylindrical hyphae, these 4-8 μm in diam., golden succineous, inside granular and becoming orange brown to reddish brown in the Melzer, rounded-obtuse at their free tips. Covering of the stipe by dermatocystidia which are ± clavate with firm walls, up to 33 μm long and 14 μm broad, rising from a thin, often scarcely developed trichodermium above the longitudinal hyphae of the stipe.

On the soil in mixed pine and oak woods, also observed in mixed beech and pine woods, also in pure pine woods, for example under *Pinus patula*.

Material studied: MEXICO: Tamaulipas, Mun. Hidalgo, Ejido Conrado Castillo, 11 XI 1987 García 5465 (F, ITCV). — Same place, 9 VI 1987 (F). — Veracruz: 3 km N of Jaltipán, 8-IX 1976, Perez Ortiz 409 (F, ITCV) — near Jalapa, NW of Lencero, 14 VII 1985 García 4764 (F) — also material from New York, USA incl. the type; New Jersey, North Carolina.

T. balloui is a very distinctive species because of the relatively short spores and the orange color of the pileus. It may be considered one of the most primitive species of the genus. There is, in Mexico, one other species with bright "reddish" pileus which was collected by G. Guzmán (16371) in a woods of leaf-shedding frondose trees with *Liquidambar* in Veracruz, Cerro de la Martinica, 10 VIII 1976 and sent to Singer (INPA). It has a much longer cylindrical stipe which is densely flocconous, and white basal mycelium. The spores are 8.2-10.5 × 3-3.7 μm , fusoid to cylindrical and the cystidia are pseudocystidia. The fruiting bodies are much more small and thin than those of *T. balloui* and have a similar trichodermium on the pileus surface. This is apparently a new species, perhaps related to *T. balloui,* but is not described separately because only little dried material was available, and the subgelatinous epicutis points to *Fistulinella* (see p.111).

2. **Tylopilus plumbeoviolaceus** (Snell) Snell, Mycologia 33: 33. 1941.

   Boletus fellcus forma *plumbeoviolaceus* Snell. Mycologia 28: 463. 1936.
   Boletus plumbeoviolaceus Snell. Mycologia 33: 32. 1941.

Pileus "dull lavender," "dark heliotrope gray," "heliotrope slate," "dark vinaceous drab," some portions soon becoming "dusky brown" or "pale brownish drab," eventually mostly "drab," the discoloration starting mostly after exposure to the sun, dry and practically non-viscid in the center, viscid in the marginal half when exposed to rains for a prolonged period and remaining shining there after it has dried out while the disc is opaque, subglabrous, then glabrous, pulvinate, eventually frequently flattened, with somewhat projecting sterile margin and the latter sometimes separating from the pileus by a circular scission, 68.152 mm broad.

Hymenophore white, then assuming a color between white and "sea shell pink," eventually "light russet vinaceous" with "sorghum brown" shades, depressed around the stipe; tubes 10-15 mm long; pores concolorous, small, less than 1mm. in diameter, usually 2-3 to 1 mm, subangular and not quite regular but not radiately elongate when old; spore print "light russet vinaceous."

Stipe "pale vinaceous drab" to "light vinaceous drab," mostly somewhat marbled with paler tints when young, becoming "light cinnamon drab," usually with a pallid inconspicuous spurious network of continuing pores on the extreme apex of the stipe running down not more than 3 mm at most, all the rest of the stipe decidedly smooth, glabrous, dry, solid, thick-ventricose, almost bulbous or bulbous when young, mostly becoming cylindric in older specimens, 40-90 × 32-56 mm.

Context white, unchanging, "light vinaceous drab" immediately under the cuticle assuming the violet pigmentation also where wounded and exposed for a long time during the perios of intensive growth, fleshy and thick, hard when young, especially in the stipe; taste very bitter; odor insignificent, not disagreeable.

Spores 9.5-14 × 3-3.8 μm, most frequently 10.2-12 × 3.3-3.5 μm, fusoid-cylindric, without suprahilar depression or with a very slight one, the broadest part in the middle or in the lower third, thin-walled and smooth, pale melleous or light golden melleous (not so golden, and not so intensively colored as in *Xanthoconium,* — young spores entirely hyaline), inamyloid or slowly and weakly pseudoamyloid; basidia 21-24 × 7.5-8.5 μm, 4-spored; cystidia 32-56 × 6-15 μm, enormously numerous on the pores as well as in the tubes, with a striking golden yellow incrustation, and also pale golden yellow contents in many cases, the attenuate or thin, ampullaceous, often acuminate though at the very tip obtusate apices hyaline and free from incrustation, the main part fusoid; trama truly bilateral, of the *Boletus*-type, the lateral stratum very loose, strongly and entirely divergent and quite hyaline; all hyphae without clamp connections.

Epicutis a highly branched and interwoven trichodermium which however bears a subhymeniform outer layer consisting of dermatocystidia, these 12-44 × 3.5-7.5 μm, ampullaceous or lanceolate, more rarely subcylindrical with obtuse to subacute to subulate, tips, with pseudoamyloid contents. Covering of the stipe consisting of some short, narrowly ventricose dermatocystidia.

Chemical reactions. — KOH on surface of pileus, bleaching to a pale isabelline; on context, negative or a very pale and sordid salmon color, at places almost bay. — NH_3 and NH_4OH negative everywhere. — H_2SO_4 on surface of pileus, bleaching to a pale isabelline; on pores brownish. — $FeSO_4$ on context, slowly and weakly reacting, becoming pale sordid gray; if KOH is added later, the reacting portion of the flesh becomes pale sordid salmon or pale sordid vinaceous, at places sordid bay (likewise slow reaction). Methylparamidophenol everywhere negative.

On earth under oaks either on bare sandy soil or among grasses, gregarious from May until fall.

The type of *B. felleus* f. *plumbeoviolaceus* Snell, the basionym, was described from New York. The species occurs also in Florida and west to Texas and Mexico; also in Honduras.

Material studied: MEXICO: Nuevo León: La Camotera (Mun. de Santiago), Valenzuela 106. — Las Adjuntas, Garcia 56 and 84. — Puerto Genovevo, García 72 (all in the Herb. Universidad autónoma de Nuevo León , Depto. Fac. Ciencias Biológicas and F). — Tamaulipas Mun. Hidalgo, El Chorrito, 30 X 1985 García 5436 (F, ITCV). — Hidalgo: road to Tamazunchale-Zimapán, Minas Viejas, 31 VII 1981, García.

3. **Tylopilus jalapensis** Sing. & Garcia spec. nov. (Pl.22, fig.13).

Pileus brown, yellowish in part, dry, convex, 45-75 mm broad.

Hymenophore tubulose, tubes pink, 8-13 mm long; pores pink to ochraceous, small (2-3 per mm), angular when mature. Spore print not obtained.

Stipe brown like the pileus, smooth or with some longitudinal lines, at the apex pale yellow, in some specimens with red dots at the base, 45-90 mm long.

Context whitish, becomes brown with a rusty tinge when bruised; odor none, taste unknown.

Spores 6.3-9.5(10) × (3)3.3-4.5(5) μm, [Q = 1.8-2.6, average (25 spores) 2.1], ovoid to fusoid thick in lower third and narrowing, sometimes subacute above, fewer thickest in the middle, hyaline to subhyaline in KOH, not pseudoamyloid with the wall 0.2-0.3 μm thick, without germ pore or apical truncation.

Hymenium: Basidia 15-21 × 5.8-6 μm, (2) 4-spored. Cheilocystidia 21-29 × 5.8-6.5 μm, ampullaceous to subcylindric, often flexuous or irregular, hyaline, thin-walled. Pseudocystidia 16-28 × 5-7.5 μm, ampullaceous, contents granular or amorphous, chestnut in the Melzer, moderately numerous. Cystidia not much differentiated from the pseudocystidia but having the internal granulation yellowish in Melzer, otherwise absent.

Hyphae: Inamyloid, without clamp connections. Hymenophoral trama now subregular (young material not available).

Cortical layers: Epicutis of pileus ochraceous brown but under immersion lens pale melleous, a deep layer of tangled, pseudoamyloid hyphae and strands of hyphae running in all directions, with walls 0.1-0.5-1 µm thick, 3-6, mostly 3-4 µm broad. Covering of stipe: Hymeniform layer or epithelium of spherocysts or short-ventricose cells or chains of such cells, these 9-14 µm diam., subhyaline but the end-cells of chains often fuscous inside; dermatocystidia present as end-cells of a chain of cells (12)28-35 × 6-13 µm, ampullaceous to ventricose.

Chemical characters: KOH on pileus negative; on context ocher; on pores brown. — NH_3 everywhere negative.

Under *Quercus*.

Material studied: MEXICO: Xalapa, near border of Veracruz: NE of Lencero, 900 m alt., 14-19 VII 1985, García no. 4767 (F, ITCV).

This species is macroscopically not fully known, but differs from the other species mainly in the vestiment of the stipe and the chemical reactions.

4. **Tylopilus mitissimus** Sing. & Gómez spec. nov. (Pl.4, fig. 14; pl. 5, fig.15).

Pileus brown violet or violet brown, when young often in part deep chocolate brown, in part livid, dried "Kis Kilim", in part, pl. 1419, even "old bronze" (M&P), velutinous, dry, with at first incurved, eventually narrowly projecting (1mm) margin, convex, 56-67 mm broad.

Hymenophore tubular, tubes white eventually pinkish, depressed around the stipe, pores concolorous with tubes, dried near "gold" M&P, on injury becoming brown, small (0.2-0.5 mm wide when dried), angular but isodiametric, on the stipe all radial pore walls short decurrent.

Stipe at first white, later becoming cream greyish, sordescent, (dried "teak wood" M&P), smooth but drying somewhat folded-rugose, and macroscopically quite glabrous, under a lens weakly velutinous but intermittently so, not flocculose-pustulate nor reticulate anywhere even at the extreme apex, solid, obclavate or ventricose, 55-71 × 20-30, at apex × 13-15 mm. Veil none. Basal mycelium white. Context white, in base of stipe, gray or blackish, unchanging; taste mild; odor slight.

Spores 9-13 × (3)3.5-(5) µm, mostly fusoid subhyaline (KOH), few pseudoamyloid, smooth. Q = 2.4-3.1.

Hymenium: Basidia clavate, 4-spored. Cystidia scattered, inamyloid, with or without contents, mostly ampullaceous or fusoid, but also clavate with 1-2 mucros at the apex, 32-50 × 6.8-11.2 µm. Pseudocystidia moderately numerous in tubes but very numerous, even crowded near pores, versiform, 22-48 × 4-11.5 µm, often ventricose, ampullaceous, fusoid, or clavate, the ampullaceous ones often granular within and with flexuous necks, contents strongly pseudoamyloid except sometimes in the neck of ampullaceous ones. Cheilocystidia 21-29 × 3.5-5.5 µm, hyaline, thin-walled, without optically visible contents, often subcylindric and flexuous, often thin appendiculate, neither wall nor interior pseudoamyloid.

Hyphae without clamp connections. Hymenophoral trama bilateral of the *Boletus*-subtype. Oleiferous hyphae 6-10 m broad, in hymenophoral trama narrower, some without, some with granular pseudoamyloid contents.

Cortical layers: Epicutis of the pileus a trichodermial palisade, consisting of a pile of subparallel erect hyphal ends which become at maturity somewhat interwoven, mostly pale yellowish gray, the rest subhyaline, with hyaline walls and intercellular dissolved pigment, the walls thin or more rarely firm, not thick, the cell chains filamentous, the terminal members obtuse or rarely subacute, 35-60 × 3.5-5.5(6.5) µm and not or scarcely different from the lower cells, not gelatinized or incrusted, not noticeably pseudoamyloid. Dermatobasidia none.

Surface of the stipe a continuous layer of subhymeniformly arranged dermatocystidia which are vesiculose, ventricose, ampullaceous, 19-53 × 4-11 µm with thin or firm wall, sometimes with a sterigma-like appendage, hyaline or with the same pigment as the elements of the epicutis of the pileus, inamyloid or pseudoamyloid as the pseudocystidia of the hymenium, some with one cross wall but all originating from the vertical hyphae of the stipe rind.

Chemical reactions: NH_3, slightly livid gray. NH_4OH: negative.

In *Quercus-Magnolia* forest, on the ground under oak.

Material studied: COSTA RICA: Cartago, La Chonta, 7 VI 1983, Singer B 14090 & L.D. Gomez (F), type.

This species is quite remarkable having initially white stipe which is entirely smooth, the practically unchanging context and a characteristic distribution of the pseudocystidial elements. It appears to be closest to *T. badiceps* (Peck) Smith & Thiers but has longer spores and different epicutis elements than those described by Wolfe (1981) on the type and Singer on D.P. Lewis 3115, from Texas. The reaction of all dermatocystidia with Melzer is weaker (only guttulae nearly pseudoamyloid), and cheilocysttidia are present and inamyloid.

5. **Tylopilus montoyae** Sing. & García spec. nov.

Pileus pinkish brown with pale vinaceous shades, tomentose, convex, + 60 mm broad.

Hymenophore tubulose, tubes and pores concolorous, whitish-pink to greyish pink, turning brown where bruised, pores 1-2 per mm and angular; tubes 8-10 mm long, adnexed, may become blackish when bruised.

Stipe concolorous with the pileus but slightly paler or yellowish white at the base when dried, macroscopically smooth and glabrous or subglabrous, under a lens finely pruinate, besides over the apical 8 mm finely (lens!) reticulate, slightly tapering upwards (to 5-6 mm), ± 110 × 9-10 mm.

Context white in the pileus, blackening when bruised; whitish to yellowish white in the stipe and pinkish brown to blackish when bruised. Odor slight, not remarkable; taste mild when dried.

Spores 10.8-12.8(14.4) × (4.2)4.6-5.6 μm, few giant spores at times reaching up to 17.5 × 5 μm, ellipsoid-fusoid, with subacute apex, some amygdaliform, most without suprahilar depression, rather pale mellous, turning orangy brown in Melzer, smooth and without germ pore or truncation.

Hymenium: Basidia 24-31.2 × 8.8-11.2 μm, 4-spored. Pseudocystidia 36-68 × 7.2-16.4 μm ventricose-fusoid to ampullaceous with amorphous or granular reddish brown content or contents more homogeneous in some, thin walled. Cheilocystidia proper not differentiated but pseudocystidia of pore region slightly shorter and much narrower, contents granular, in Melzer reddish brown, some cylindrical, abundant.

Cortical layers: Epicutis of the pileus a trichodermial palisade, with the terminal cells 28-50 × 3.2-9.6 and often cystidioid, much like hymenial pseudocystidia but more variable in shape and some with yellowish-hyaline contents, others with irregularly distributed orangy brown (in Melzer) contents, all without clamp at basal septum.

Chemical color reactions: KOH on pileus orange yellow; hymenophore yellow; context yellow. — NH_4OH everywhere turning yellow; also with NH_3 vapors.

Solitary in mesophilous montane forest in June.

Material studied: MEXICO: Veracruz: Road San Andrés-Tlalnehuayocan, Plan de Sedeño. 28 VI 1986, Leticia Montoya-Bello 699 (F, type; XAL).

6. **Tylopilus alboater** (Schwein.) Murr., Mycologia 1: 16. 1909. (Pl. 4, fig. 17; pl. 5, fig. 16; Pl. 6, fig. 18, 18A, 19,20A-B).

Boletus alboater Schweinitz, Schr. Naturf. Ges. Leipzig 1: 95. 1882.
Boletus nigrellus Peck, Ann. Rep. N.Y. State Mus. 29: 44. 1878.
Suillus alboater Kuntze, Rev. Gen. Pl. 3(2): 535. 1898
Suillus nigrellus Kuntze, Rev. Gen. Pl. 3(2): 535. 1898.
Porphyrellus alboater Gilbert, Bolets, p. 99. 1931.
Porphyrellus nigrellus Gilbert, Bolets, p. 99. 1931.

Pileus very dark brown to nearly black with a violet-greyish bloom, or dark smoky drab to brownish grey, often with a flesh colored tint, distinctly velutinous, dry, with at first incurved then narrowly projecting sterile margin, convex becoming nearly flat or irregular in age 60-113(230) mm broad.

Hymenophore tubulose, white to pale grey, later flesh colored, when injured changing slowly to black or reaching blackish though pink or purplish brown, adnate to depressed with decurrent lines; tubes 5-12 mm long; pores somewhat variable in shape and folded together, later mostly rounded-angular, concolorous with the tubes and changing as tubes when bruised or touched. Spore print deep rosy salmon to brownish-flesh color.

Stipe concolorous with the pileus but pale at the apex at least when young, usually reticulated above, the longitudinal veins stronger than the cross veins, more rarely scarcely reticulate even at the apex, otherwise pruinose to velvety, blackening when rubbed, usually heavy and often irregular, or slightly tapering upwards, rarely tapering downwards, solid, 32-105 × 14-29 mm; basal mycelium sordid white; veil none.

Context white to creamy grey or pale brown, changing to pinkish or purplish brown bruised, eventually becoming blackish, firm in the stipe, thick in the middle of the pileus; taste mild, sometimes nutty; odor none or not remarkable and weak.

Spores (8)9-13 × 4-4.8(5) μm oboclavate or oblong to fusoid without suprahilar depression or applanation, often broadest near lower end, pale brownish melleous, rather thin-walled, but sometimes a few moderately thick-walled, some pseudoamyloid, smooth, without germ pore or truncation.

Hymenium: Basidia (24)29-41 × 7.3-11.2 μm, 4-spored. Pseudocystidia (32)60-65 × (8)11-12 μm constantly fusoid-ampullaceous, brown, the neck up to 34 μm long and 3.5-4.4(6.5) μm broad; some cystidia without contents; cheilocystidia little differentiated.

Hyphae without clamp connections, brown in the mediostratum hyaline and loosely arranged in the lateral stratum: *Boletus*-type.

Cortical layers: Epicutis of the pileus made up of a palisade of hairlike hyphae which are dark fuscous, slightly attenuated upwards, but obtuse at the tip, rarely subacute, brittle, often thickish walled (wall 0.5-0.9 μm thick), densely arranged, 4-7.5(12) μm broad. Similar hairs also found on the covering layer of the stipe. The palisadic structure eventually disorganized and approaching a trichodermium.

Chemical color reactions: KOH on pileus black, on tubes dark brown, or reddish brown; on stipe reddish to vinaceous brown, on context wine red. — NH_4OH on pileus reddish brown, on tubes dark brown, on stipe blackish, on context caesious. — $FeSO_4$ on stipe slightly olive; on context slowly slightly grey.

In mixed woods, often in low places, near deciduous trees, in Mexico in mountain woods on the soil.

Material studied: MEXICO, Mexico, D.F.: road to Villa del Carbón, 2400 m alt., Gonzales-Velazquez 157 (F, ENCB). — Also material from USA (Ohio, North Carolina, Florida).

This species has also been indicated from Malaya, and Singapore (Corner 1972) but this refers to *T. pachycephalus* (Mass.) Sing. (Singer 1986). The true *T. alboater* has not been observed south of Mexico.

7. **Tylopilus obscurus** Halling, Mycotaxon 34: 109. 1989. (Pl.7, fig.21-22).

   Tylopilus montanus Sing., Fieldiana N.S. no. 21. 1989.

Pileus fresh chocolate brown with purplish hue when young, dried butterscotch, later between Vandyke brown (and "cocoa brown M&B), old with even darker areas, with easily peeling cuticle, but dry, velutinous and unshining, tending to be minutely rimulose, then often grossly cracked, eventually subglabrescent, the margin at first acute, later obtuse, at first narrowly incurved, surface pulvinate, then convex with mostly applanate or narrowly depressed center, up to 111 mm broad.

Hymenophore tubulose. Tubes, moderately long, adnexed then depressed around the stipe pores and tubes white to pale cream, becoming fumosous and cinnamon stained, subisodiametric, round to angular and small (0.8 mm) at first but varying somewhat angular at last and 0.5-1 mm wide. Spore print slightly deeper than "roe" (M&P).

Stipe concolorous with pileus, base slightly lighter colored, at first apex light cinnamon, narrowly reticulate, later developing a strong rib-like sulcation which in the middle portion becomes coarsely (almost alveolate) reticulate, are fine and disappearing somewhere in the lower third where the stipe becomes totally smooth to (mear middle zone) finely, sometimes indistinctly reticulate, the reticulum concolorous with the ground color of darker soil ± bulbous but soon in upper portion more equal, 81-90 × 41-51 mm (apex at least × 30) mm. Veil none. Basal mycelium "maple" M&P.

Context more or less whitish or darkening to ocher; no discoloration observed (but evident in hymenophore). Taste mild, eventually often like starch.

Spores 10-14.8 × 4-5.5(6) μm, most frequently 11-13 × 4-5 μm, rather variable in shape (but Q not reaching 3), most frequently fusoid to cylindric or ellipsoid to oblong, few constricted or tilda shaped, with thin or firm to thickish walls, with a slight suprahilar depression or applanation, subhyaline or palest melleous, with homogeneous wall and smooth, slowly and very weakly to distinctly but moderately strongly pseudoamyloid, without germ pore.

Hymenium: Basidia 21-35 × 8-10 μm, 4-spored. Cystidia and pseudocystidia present, the latter with inside corpuscules which are badious in Melzer 30-43 × 3.5-7 μm, ± ampullaceous with apex about 3 μm. Cheilocystidia not differentiated.

Hyphae rather narrow without clamp connections. Hymenophoral trama, pale yellowish in the mediostratum, hyaline in the lateral stratum, of the *Boletus*-subtype.

Cortical layers: Epicutis a trichodermial palisade of thin to firm walled hyphae, the terminal members obtuse or subacute, some hyaline but most yellowish brown inside, but wall likewise yellowish brown or hyaline, 14-30 × 3.5-7(10) μm. Hypodermium a cutis of more deeply chestnut colored elongated hyphae. Pigment intraparietal or interparietal and perhaps occasionally weakly incrusting. The epicutis cells with pseudoamyloid (chestnut) interior bodies. Covering layer of the stipe in the reticulate zone formed by a subhymeniform layer consisting of dermatocystidia. These 15-36 × 3-7 μm, hymeniformly arranged, hyaline or yellow brown or succineous, much like the dermatocystidia of the pileus, rising from a trichodermium of both short and elongated cells (the short ones e.g. 15 × 11 μm) whereby the dermatocystidia form the terminal cells of the trichodermial cells, but the latter are basically hyphous. Total depth of the covering layer: 70 μm.

In mixed oak woods *(Querceto-Magnolietum)* under oak on the ground.

Material studied: COSTA RICA: Cartago: La Chonta, 11 VII 1982, ca. 1500 m alt. L.D. Gomez no. 18199 (F) type. — Also Colombia: Antioquia, Guaune, ± 14 km E of Medellín, 2850 m, Halling 5008 (NY) isotype.

Our description covers only the material from Costa Rica which is obviously a mature, but younger stage of the Colombian material. Singer recognized this *("T. montanus")* since 1986 as an autonomous species. Halling's (1989) opinion, assuming association with oak *(Q. humboldtii)* in Colombia, is sustained by the fact that the Costa Rican material is associated with *Quercus,* presumably by ectomycorrhiza.

8. **Tylopilus nicaraguensis** Sing. & Ivory, Beih. Nov. Hedw. 77: 107. 1983.

Pileus buff to vinaceous buff fresh (according to collector), glabrous, smooth with somewhat papery cuticle, soon becoming flat, about 100 mm broad.

Hymenophore tubular, tubes vinaceous buff, easily separating from the context of the pileus, rather long (e. gr. 7 mm), adnexed to the stipe-apex, pores concolorous, apparently unchanging, round or somewhat gyrose, 1 mm wide.

Stipe vinaceous-buff with sepia base, slightly reticulate when fresh, only slightly subreticulate-striate when dried (under a lens), woolly at apex, otherwise smooth and subtomentose, solid, subequal or subventricose, ± 80 × 20 mm. Basal mycelium whitish.

Context white, darkened to melleous in stipe. Taste very bitter. Odor musty.

Spores 8-10.3(13.8) × 3-3.7(4) µm, fusoid to cylindrical, hyaline to subhyaline in KOH, appearing inamyloid (yellow-melleous) in Melzer's reagent when seen singly, but in accumulations very slightly pseudoamyloid, smooth, without germ pore.

Hymenium: Basidia 22 × 9.3 µm, 4-spored. Cystidia numerous on pores, less so in the interior of the tubes, 20-35 × 3.36 µm, most frequently narrowly ampullaceous or almost subulate but also often narrowly utriform or even clavate, hyaline and thin-walled but some thinly colored by a pale melleous strictly applicate and an entire resinaceous incrustation both turning reddish-brown in Melzer, the apex (of ampullaceous cystidia) 1.5-3.3 µm across and obtuse, hyaline and optically empty inside.

Hyphae of the context hyaline, thin-walled, inamyloid, 2.7-5 µm broad. Hymenophoral trama bilateral of the *Boletus*-type, the mediostratum yellowish and consisting of 1.3-4 µm broad filamentous hyphae, the lateral stratum hyaline and gelatinized, of recurved hyphae 2-8 µm broad.

Covering layers: Epicutis of the pileus a very slightly and partially gelatinized cutis, consisting of 0.4-4.3 µm broad hyphae which are either repent or, here and there, curved upwards to form an obsolete trichodermial upper layer, most quite hyaline but some with the same applicate resinous incrustation as the hymenial cystidia. Hypodermium not or scarcely gelatinized, of hyphae similar to those of the epicutis and 1.3-11 μm broad. Covering of the stipe almost hymeniform, consisting of dermatocystidia about 24.5 × 5.7 with about 2.7 μm broad neck, much like the ampullaceous hymenial cystidia.

The carpophores readily parasitized by a *Sepedonium*.

Solitary under *Pinus caribaea,* with dense ground cover of herbs and various woody shrubs.

Material studied: NICARAGUA: Slima Sia, V 1976, Ivory S/20 (F), typus.

This species differs from *T. neofelleus* Hongo in larger spores, from *B. griseipurpureus* Corner in somewhat narrower spores and pores (but the description of the taste is not given), from *T. minor* Sing. in being much larger and having woolly stipe apex, having smaller, at the edges of the tube-walls more incrusted cystidia and different mycorrhizal association; *T. areolatus* Hongo seems closest and somewhat nearer to *T. minor* Sing. than to *T. nicaraguensis* which has narrow spores, non-areolate pileus, woolly apex, narrower cystidia, darker stipe, but is associated to pine as is *T. areolatus.*

9. **Tylopilus sanctaerosae** Sing., Beih. Nov. Hedw. 77: 112. 1983, fig. 39 (p. 332).

Pileus with strands of thin felt fuliginous on livid greyish ground which becomes visible in mature caps, dry, smooth, with the margin not or scarcely projecting, convex, at maturity slightly depressed in the center, reaching 92 mm broad.

Hymenophore tubular, tubes pallid, then pink, only slightly depressed around the stipe, and very slightly lamellar there; pores concolorous with tubes, dirty grayish pinkish then soon blackening when touched, up to 1 mm in diameter.

Stipe concolorous with the pileus but less tomentose, finely reticulated concolorously at the apex, below that minutely scurfy-subtomentose, macroscopically smooth but under a lens scurfy-subpustulate, the transition from the reticulated apex to the non-reticulated portion of the stipe being gradual, not abrupt, solid, broadest in the lower third, slightly narrower and equal in the upper part, tapering to an obtuse, non-radicant base below, 85 × ± 15 mm × 17 mm (apex). Veil none. Basal mycelium rather scanty, white.

Context white, when bruised at first dirty grayish-pinkish, soon blackening. Taste mild. Odor weak, not characteristic.

Spores 10-13 × 3-4.3 μm, Q (2.5)2.7-3.5(3.9), subhyaline to palest melleous-hyaline, fusoid few cylindrical, smooth, the wall firm and homogeneous, about half the spores weakly pseudoamyloid especially in superposition, the rest inamyloid, without germ pore or callus.

Hymenium: Basidia 14-35 × 7-9.5 μm, mostly 23-24 μm long, (2-) 4-spored. Cheilostydia 21-13 × 4-9.5 μm, versiform, most frequently ventricose, a few with one or two finger-like appendages at the tip, often brown (KOH) in dried, mature material especially where the pores had been manipulated, others hyaline. Cystidia sparse, towards edge moderately numerous, 20-45 × 5.5-10.5 μm, ampullaceous, thin to rarely firm-walled, hyaline, few brownish inside (KOH), but some opaque, the obtuse apex 3-4 μm across, tapering upwards or cylindric, some (pseudocystidia) with amorphous filiform inside bodies, but inamyloid.

Hyphae: Hymenophoral trama basically hyaline and of the *Boletus*-subtype, but in dried material with some intracellular and/or incrusting spadiceous pigment, hyphae often opaque with firm walls, often with spherocysts at the end of the chains, these usually internally brownish ocher, end cells often cystidia-like and (as endocystidia) e. g. 16-33 × 10-13 μm. All hyphae inamyloid and without clamp connections.

Cortical strata: Epicutis of the pileus (the fascicles of hyphae forming the tomentose upper layer) consisting of erect to ascending hyphae with the terminal cells cystidiform, these dermatocystidia mostly not incrusted by pigment but with a distinct intraparietal spadiceous pigment, 23-70 × 5.5-9.5 μm, mostly ampullaceous or at least tapering upwards, fewer utriform or cylindrical, rising from a trichodermium of hyphous structure, in the uppermost layer generally depressed into a pseudocutis, the cells cylindric with hyaline to pale fuscidulous walls but frequently pigment is incrusted by spadiceous pigment; some rare cells sperocyst-like. Covering of the stipe consisting of hymeniform layers of ventricose to clavate cells 23-44 × 7-13 μm, these dermatocystidia or dermatopseudoparaphyses basidiole-shaped and either hyaline or spadiceous and then incrusted or pigment confined to walls (intraparietal), all dermatocystidia rising from a shallow, pigmented trichodermium superimposed on the mostly vertical surface hyphae of the stipe rind. Dermatobasidia absent.

Chemical characters: NH_4OH — negative.

On the ground under *Quercus oleoides.*

Material studied: COSTA RICA: Guanacaste: Parque Nacional de Santa Rosa. 25 VII 1981. L.D. Gomez and Rolf Singer B 12436 (F) - typus.

This differs from *T. nigerrimus* Heim in much narrower spores, absence of large ventricose dermatocystidia on the pustules of the stipe and smaller hymenial cystidia.

T. subniger differs from *T. sanctaerosae* not only by the key, p. 10-11) characters but also in the percentage of pseudocystidia in the hymenium, and the presence of a spadiceous, incrusting pigment in the trichodermia, whereas *T. sanctaerosae* has intracellular pigment.

10. **Tylopilus guanacastensis** Sing. Beih. Nov. Hedw. 77: 110. 1983. (Pl. 7, fig. 23).

Pileus dark burnt umber or brown, yellowish brown, dried between "racquet" and "London smoke", dry, subvelutinous to velutinous, eventually often scabrous on chamois ground, often with some superficial fibrils becoming visible under a lens, convex often applanate or subumbonate on the center but rarely depressed there, 50-90 mm broad.

Hymenophore tubular, tubes pale chocolate brown, khaki, or olive drab, when bruised turning deep chocolate brown, depressed around the stipe, 9-11(15) mm long; pores concolorous with the tubes and equally discolored when bruised, small (less than 1 mm wide).

Stipe sienna, raw umber, or coffee brown, even livid maroon, up to one third of its length reticulate by a fine network of veins on slightly lighter ground, below that rarely also reticulate, mostly smooth and covered by a densely almost furfuraceous-dotted, small-pustulate pruina, solid, mostly more or less cylindric, 45-58 × 10-14(18) mm; veil none; basal mycelium white, in places tending to pale brownish.

Context white, unchanged when bruised, in dried condition mostly dirty pallid; taste mild; odor insignificant.

Spores 9-13(14) × 3-4.5(4.8) μm, most frequently 10-12 × 3.5-4.2 μm (Q = 2.2-3.1), fusoid, cylindric, subhyaline (KOH) or pale buff, inamyloid but a variable number (few or the majority) pseudoamyloid.

Hymenium: Basidia 21-38 × (6.5)9.5-11.5 μm, 4-spored. Cystidia moderately numerous, without contents and inamyloid, mostly ampullaceous 25-48 × 9-10 μm. Pseudocystidia in tube interior scattered, toward pores numerous, pale melleous-brownish with pseudoamyloid, coarsely granular interior, versiform, (19)25-44 × 6-14 μm, in both cystidia and pseudocystidia the wall hyaline and thin. Cheilocystidia not differentiated.

Hyphae hyaline, inamyloid without clamp connections. Hymenophoral trama with distinctly diverging hyphae in the lateral stratum (apparently of the *Boletus*-type).

Cortical layers: Epicutis of pileus a trichodermium consisting of elongated, non-incrusted hyphae but the uppermost cells often applicate (a "pseudocutis"), with brownish inclosures which are pseudoamyloid becoming deep chestnut brown in Melzer; no subisodiametric or spherocystoid bodies and no incrustation present. Hypodermium more cutis-like. Covering of the stipe on often longitudinally seriate pustules which are formed by a hymeniform layer of basidioles, 2-spored basidia and dermatocystidia, the latter 17-35 × 7-10 μm, in shape much like the hymenial cystidia and dermatocystidia, most with resinous cap.

Under *Quercus oleoides* at up to 80 m alt. on sandy soil, fruiting from August till November.

Material studied: COSTA RICA: Guanacaste, 7 km NW of Bagaces, X 1982, L.D. Gomez 10724 (12725?), F, type. — Gomez & Alfaro, VIII 1983, 21316 (F). — Gomez & Alfaro 22098, 29 XI 1983 (F), from the same locality has some features reminiscent of *T. montoyae* (no. 5).

The species is evidently closely allied to *Tylopilus subpunctipes* (Peck) Smith & Thiers from it differs in non-scrobiculate pileus, unchanging context and a much deeper colored stipe. Also the cystidial contents never changes to vinaceous red or reddish, as indicated by Smith & Thiers (1971).

11. **Tylopilus lividobrunneus** Sing., Sydowia, Beiheft 7: 103. 1973. (Pl. 7, fig. 24-25).

Pileus livid, then light greyish-brown, glabrous or subglabrous, dry, convex, obtuse, 55-60(150) mm broad.

Hymenophore tubular; tubes pinkish, depressed around the stipe, pores small (up to 1 mm wide), concolorous, not browning.

Stipe dark brown, slightly ribbed but not reticulate, glabrous or subglabrous, under a lens very slightly pubesent, equal but with narrowed apex, mostly longer than 100 mm, about 15(30) mm broad, at apex 10 mm broad.

Context white, unchanging. Odor insignificant or nil. Taste mild.

Spores 10-13 × 4-6 μm, mostly 10-12 × 5-6 μm, fusoid, oblong, clavate or ellipsoid, mostly ± fusoid, smooth.

Hymenium: Basidia around 18.5 × 8 μm, 4-spored. Pseudocystidia numerous, with coarse round and filamentous inside bodies, which are pseudoamyloid, (22)40-52 × 6-13 μm, ampullaceous with up to 22 × 2-3.5 μm neck and thin wall. Cystidia few or absent. Cheilocystidia differentiated, 19-26 × 6-8 μm, without contents and not pseudoamyloid, thin-walled, hyaline, ampullaceous or fusoid.

Hyphae inamyloid, without clamp connections. Hymenophoral trama of the *Boletus*-subtype, mediostratum yellowish, lateral stratum hyaline.

Cortical layers: Epicutis of the pileus a trichodermium (not a palisade), many terminal members like the pseudocystidia of the hymenium (rusty brown in Melzer), others like the cheilocystidia (most pseudocystidioid), all clavate, ampullaceous or fusoid, 24-55 × 2.5-12 μm, some of these as well as the lower cells with slightly thickened wall, the lower cells elongated, 1.4-7(12) μm broad, often incrusted by yellowish to gold brown (pseudoamyloid) pigment. Covering of stipe consisting of a continuous hymeniform layer of dermatopseudocystidia, these either large, 18-35 × 6-14 μm with strongly pseudoamyloid inside bodies, or small 13-32 × 3-6 μm, hyaline in KOH, with scattered pseudoamyloid inside-bodies, both types versiform (clavate, ampullaceous, some subcylindrical or ventricose), few with 1-2 secondary septa.

Chemical characters: NH_4OH on surface of pileus negative; on hymenophore yellowish.

Under *Quercus* (not *Q. oleoides)* on the ground, at an altitude of 1850-2300 m.

Material studied: MEXICO: Oaxaca, San Agostín, 10 VII 1969, Singer M8399 (F), Typus. — COSTA RICA: Cartago, Interamerican Highway km 56, L.D. Gomez 37 (F).

This species seems rare in Costa Rica. It is similar to *T. gomezii* Sing. but does not change color when bruised.

12. **Tylopilus gomezii** Sing. sp. n. (Pl. 8, fig. 26).

Pileus light umber or umber, mostly on chamois ground, dried between "teak wood" and "Madrid" the marginal zone between "alamo" and 13 G 6, finely scurfy-rivulose, at least visible under a lens, smooth with narrowly projecting sterile margin, convex, obtuse, 84-146 mm broad.

Hymenophore tubular; tubes sordid pinkish gray when mature, to 13-17 mm long, depressed around the stipe; pores concolorous, brownish when bruised, small (0.3-9 mm diam.), round or angular, not or very slightly lamellar around the stipe.

Stipe leather brown or chocolate brown, only with ridges which occasionally anastomose to form an incomplete network at the very apex of the stipe, or with a reticulation over the top (1)7 mm of the stipe, entirely and densely (or in ridges) pustulose all over, these ridges gray-brown in dried condition, solid, always ± obclavate and 115-167 mm long, 9-14 mm broad at apex, 13-27 mm broad at base. Veil none; basal mycelium white.

Context white, becoming pale rose to dirty purplish when bruised. Taste mild, with bitterish aftertaste. Odor slightly soapy.

Spores (8)10-13(14) × 4-5.5(6.5) μm, Q = 1.8-2.8, mostly 2-2.7) oblong to ellipsoid or mostly fusoid, smooth without germ pore, with or without suprahilar depression or applanation, melleous-subhyaline, inamyloid or very weakly pseudoamyloid.

Hymenium: Basidia 4-spored. Cystidia numerous both as pseudocystidia and cystidia the former more numerous, especially at pores, 39-55 × 7-11 μm, ampullaceous with the neck 2-3 μm wide, subhyaline to light chestnut in KOH but with coarse pseudoamyloid internal granulation, the latter without colored or pseudoamyloid contents and rather rare at pores and inside tubes, always fewer than pseudocystidia. Cheilocystidia not or scarcely differentiated.

Hyphae generally without clamp connections, inamyloid, some indistinctly pseudoamyloid in dried material (oleiferous hyphae); hymenophoral trama of the *Boletus*-type.

Cortical layers: Epicutis of the pileus (corresponding to the superficial fibrillose-scurfy covering) yellowish to chestnut or in parts hyaline, a trichodermium of strongly interwoven hyphae, almost 70-80 μm deep, these not gelatinized, its cells elongated and 3-7.5 μm broad, with or without intracellular melleous contents, some or many with subhyaline to yellowish to pale chestnut incrustation, at the tip obtuse, often with resinaceous cap, with thin (0.2-0.4 μm) to somewhat thickish wall of 0.5(-1) μm diam., in the Melzer some globules and other inside bodies as well as thin incrustations in some of the cells becoming reddish chestnut (pseudoamyloid) or remaining just yellow (inamyloid), the pseudoamyloid elements clustered in some areas of the epicutis, rarer in other areas; some terminal cells cystidioid, dermatocystidia e. gr. 30 × 5 μm, pseudoamyloid, hypodermium of interwoven hyphae forming some sort of a cutis with some granular pseudoamyloid incrustation visible.

Covering of the stipe (forming the pustules) consisting of dermatocystidia (most of which are pseudocystidia with yellowish granular contents in KOH), (1) like hymenial cystidia, ampullaceous 41-62 × 8.5-11.5 μm, mostly clavate.

Chemical color reactions: KOH or pileus chestnut to orange, on pores yellow, on stipe deep brown, on context ochraceous. — NH_4OH on pileus negative; on pores yellow, or stipe a deeper brown, on context pale ochraceous.

On the ground in montane mixed oak or oak-*Magnolia* woods with or without Lauraceae (not under *Quercus oleoides),* late-fruiting (August to November); in Mexico in *Quercus-Alnus* woods.

Material studied: COSTA RICA: Cartago, 2 km SE of junction of Highway No. 2 to San Cristobal at 1800 m alt., XI 1983, Gomez & Alfaro, No. 22018 b. (F) type — Puntarenas, Upper Río Burú & Rio Bellavista. August 1983, between 1700 and 2500 m alt., Gomez et al. 21453. (F). MEXICO: Querétaro, road San Juán del Río — Jalpán km 149,

1950 m alt. 12 VII 1985, Valenzuela 4696, det. Singer & García (F, ITCV).

This species is very similar to *T. lividobrunneus* as far as habit and spores are concerned. The three collections differ sharply from the latter because of the distinct discoloration observed on bruising of the various parts of the carpophores.

13. **Tylopilus subcellulosus** Sing. & García spec. nov. (Pl. 8, fig. 27).

Pileus brown, soon brown rivulose-rimulose all over on cream colored ground, dry, with not projecting margin ± convex, 75-100 mm broad.

Hymenophore tubulose, tubes pink, ± 11 mm long depressed around the stipe, pores concolorous with tubes, small (0.5-0.8 mm diam.) angular when mature. Spore print not obtained.

Stipe brown to slightly tinged violet, at base cream color, seemingly smooth but under a lens finely pustulate and the pustules often arranged in reticulate pattern, but not venose-reticulate, subbulbous below with narrowed (10-15 mm across) apex, 110-115 mm long.

Context white and unchanging when bruised, about 15 mm thick. Odor none or somewhat fungoid. Taste bitter.

Spores 9.5-13.2 × 3.5-4.2 μm, mostly 10.5-11.5 × ± 3.5 μm, with a few giant spores to 18 μm long and 3.5-5 μm broad, the majoritiy inamyloid, but some pdeusoamyloid either only in the interior or also in the walls, strictly fusoid, quite smooth, without germ pore and apical truncation, subhyaline in KOH.

Hymenium: Basidia 15-26 × (6.2)7.5-8(10.2) μm, (2)-4-spored. Cystidia and pseudocystidia of the same shape, both on edge and tube interior, 20-40 × 6.7-12.5 μm, rather thin-walled, clavate-ventricose, either with a long (up to 20 μm) and subacute appendiculus, and with or without internal guttate pseudoamyloid bodies (one or several), or else with a slight inamyloid or pseudoamyloid incrustation, or at least a very slight applicate incrustation.

Hyphae hyaline, a few finely brownish incrusted, especially in hypodermium and at pore level, the incrustation often pseudoamyloid, not strongly gelatinized except in the lateral stratum of the hymenophoral trama, the latter bilateral of the *Boletus*-type; oleiferous hyphae present.

Epicutis — a layer of 100-300 μm deep trichodermium which may in the upper zone be organized as a palisade or subhymeniformly and the terminal members variable in shape from cellular-spherocystoid to cystidiform or just cylindrical, often with a narrowed neck or appendix or ventricose-mucronate, the elongated ones 31-45 × 5-12 μm, the short or isodismetric ones 12.5-35 × 10-17 μm, often as globose as 13 × 13 × 13 μm; the lower cells of the chains are either elongated and 5-12 μm broad, or a few spherocystoid ones intercalated (e.g. 12 μm) or at least some short-ventricose (e.g. 16 × 10 μm) ones present among the cylindrical ones, all neither gelatinized nor incrusted, with intracellular pale fuscidulous pigment. Hypodermium reduced to a few tiers of hyphae forming a cutis. Covering of the stipe devoid of spherocysts.

On the ground in mixed woods *(Quercus, Magnolia, Liquidambar, Acer,* etc.).

Material studied: MEXICO: Tamaulipas, Rancho El Cielo, (Mun. Gomez Farías), 17 VII 1987. García 5451 (F, ITCV). — Rancho El Cielo, Mun. Gomez Farías, 9 VII 1984. García 3978 (F, ITCV). — Same place. García 3781 (F, type; ITCV).

This species has a rather characteristic epicutis. With its bitter taste and unchanging colors it reminds one of the scarcely violaceous forms of *T. williamsii* which however have slightly smaller spores (no giant spores observed) and even more strikingly pseudoamyloid, larger cystidia, and virtually all hyphous trichodermium. *Tylopilus vinosobrunneus* Hongo (Sydowia Beih. 8: 193. 1979) is very similar and related, but differs in the vinaceous brown pileus which is smaller (30-70 mm broad), but has relatively somewhat broader spores, and rosascent context and browning pores.

14. **Tylopilus williamsii** Sing. & García spec. nov. (Pl. 9, fig. 30).

Pileus pinkish gray, dried "oak wood": M&P, with paler or whitish margin, or varying to pale chocolate near no. 75, 1.y br (ISCC-NBS Color Charts), often more purple violet when young and dried "Cattail" M&P, dry, subtomentose, under a lens very finely fibrillose or punctate, almost velutinous in places, eventually often very finely rivulose, convex to somewhat irregular, smooth, with the margin subacute (50)-78-(90) mm broad.

Hymenophore tubular, tubes pallid to pinkish, in very young ones whitish, 10-13 mm long, adnexed or depressed around the stipe; pores concolorous with tubes becoming ochraceous or yellowish when touched, 0.5-1 mm wide. Spore print pink.

Stipe white or whitish or somewhat yellowish, brown in some parts when mature, smooth and glabrous, not reticulated but often innately fibrillose when older, solid, cylindric or more rarely slightly enlarged in the middle, rounded at base, (45)50-60 × 10-13 mm, rarely somewhat broader. Veil none. Basal mycelium scanty.

Context white, in the pileus to 19 mm thick and unchanging, in the stipe sometimes changing to (greenish-) yellowish. Taste decidedly bitter. Odor insignificant.

Spores 7.5-11 × 3.5-4.5 μm, mostly 8-9 μm long (Q = 1.9-2.7, average 2.2), more or less fusoid but often broadest in lower part, smooth, melleous hyaline in KOH, pale livid in the Melzer, but the weak reaction disappearing after a while, or in a few spores becoming distinctly pseudoamyloid in contents, inamyloid in wall, without germ pore or truncation.

Hymenium: Basidia 25-27(33) × 7-7.5(9) μm, 4-spored. Cystidia rather few, 27-48 × 6-9 μm, hyaline and without contents, inamyloid, ampullaceous. Pseudocystidia numerous on pores and in the interior of the tubes, 24-48 × 4.5-9(12.5) μm, more often ampullaceous in the tubes than on the pores, there often clavate, always with an amorphous or guttulate inside body which becomes reddish brown or rusty brown in the Melzer (pseudoamyloid), sometimes weakly so. Cheilocystidia little differentiated. All cystidial bodies may sometimes show some pseudoamyloid incrustation.

Hyphae inamyloid, without clamp connections. Oleiferous hyphae 2-7 μm broad. Hymenophoral trama distinctly of the *Boletus*-type.

Cortical layers: Epicutis of the mature pileus forming a trichodermium, with the terminal cells strongly elongated and at first often partly parallel but not consistently so and even then not truly palisadic, 2.5-6 μm diam, rarely as narrow as 1 μm, with pseudoamyloid contents, few with a (in KOH) ochraceous globule, sometimes ± claviculate or somehow widening above, but not cystidioid, with rounded-obtuse ends, the lower cells also trichodermatoid, elongated, 1-5 μm broad, brown in ×40 objective, pigment intraparietal in the outer layer of the wall. Covering of the stipe subhymeniform, elements of the hymenium 9-19(26) × 5-8(11) μm, dense but often not parallel with each other, most with golden-melleous (in KOH), pseudoamyloid (red-brown in Melzer) granular contents; dermatobasidia similar, e.g. 19 × 10 μm, 2-4-spored; dermatocystidia 30-60 × 3.5-8.2 μm, ampullaceous, more rarely fusoid-mucronate, hyaline, some (dermatopseudocystidia) with pseudoamyloid, granular contents, pedicellate, thin-walled. Also, some long hyphae protruding, rising from the inamyloid dermatocystidia or the basal trama, 2.5-3.5 μm broad, some of these with in Melzer golden-ochraceous, loose granules inside; in young specimens these hyphous protuberances are numerous and long, with obtuse tip, later they become rather rare. Some of the basic elements of the subhymeniform layer in young specimens are smaller than indicated above and sometimes thick and complex-walled.

Chemical characters: KOH on fresh pileus surface: orangy yellow, no reaction on old or dried material; no reaction on context; on pores ocher. — NH_4OH on pileus orange, on pores and context negative. — $FeSO_4$ on context greenish.

Under *Quercus* on the ground in pure and mixed, often deteriorated stands.

Material studied: MEXICO, Veracruz: 5 km NE of Lencero, road Xalapa - Veracruz 14 VII 1985, García 4759 (F, type; ITCV). — Also material from USA (Florida, F).

As compared with *T. plumbeoviolaceus,* this species has slightly smaller pileus and spores and a not truly palisadic upper layer of the epicutis with the terminal cells not cystidiform. The violet tinge of the pileus, if at all produced, disappears in the earlier development of the carpophores in *T. williamsii,* and the stipe is never livid-violet marbled as in *T. plumbeoviolaceus.* Also the chemical color reactions seem to be slightly different.

T. jalapensis is also similar and apparently related but its context, when bruised, becomes rusty brown and the elements of the stipe covering are broader, often spherocystoid. *T. vinosobrunneus* Hongo (1979) is also related but the covering layers are quite different in the Japanese species.

This species is named for Robert S. Williams who collected and annotated it first.

15. **Tylopilus costaricensis** Sing. & Gómez n. sp. (Pl. 9, fig. 28).

Pileus purplish brown, dried between "cocoa" to "Mohawk" and "leafmold," (M&P) also (20626) violaceous brown (dried "cocoa brown" with "butterscotch" margin, or "cattail" with "olive wood" margin), subtomentose, in places finely areolate to rimulose (the areolae deeper colored than background, or somewhat scurfy (20626) with rather thin and acute margin, not incurved, convex, obtuse, soon in the middle depressed, entirely concave in the end, (57)-81 mm broad.

Hymenophore tubular; tubes whitish, soon somewhat depressed around the stipe, about 7 mm long or slightly longer; pores white, soon becoming greyish brown (dried "burnt umber"), when fresh bruising brown, 0.3-0.6 or up to 1 mm wide, isodiametric or subisodiametric (somewhat irregular) with a wide zone of sublamellarly extended pores around the stipe (or without such a zone in 20626).

Stipe with gray ornamentation on white ground, often with somewhat rosy tint at apex (20626), dried brown, finely reticulated in the upper part, up to one third of the length of stipe (20621) or only a few mm's at extreme apex (20626), below smooth but beset with fine pustules sometimes both zones or either one appearing as if overgrown by fine fibrils (s.l.), solid, obclavate but with a more (type) or less (20626) developed basal appendage (as *Boletus appendiculatus),* sometimes eccentric or slightly curved, but more frequently central and straight, about (65) 81 × (21-22)81 mm; apex of stipe (10)16-17 mm across.

Context white or whitish when fresh or dried, any discoloration after bruising not indicated (20621) or browning (20626). Taste mild; odor none.

Spores 7.5-10.3(12) × 3-4(4.5) μm, in some (20626) 7-11(14) × 3.2-5 μm (Q=(2)2.2-3.4(3.6) or (20626) 1.9-3), fusoid, some obclavate, very few utriform with strangulation, most with a slight suprahilar depression, smooth, inamyloid, a few pseudoamyloid, without germ pore, with rather thin ($\pm$ 0.3 μm) wall.

Hymenium: Basidia 20-25 × 8.5-10 μm, 4-spored. Pseudocystidia on the sides and in tube interior, more crowded on the pores and there more pseudocystidia than cystidia, and mostly more ventricose-fusoid than ampullaceous, obtuse and strongly pseudoamyloid with internal granules. Cystidia more common inside the tubes, similar in shape, but without visible contents, these and pseudocystidia 23-42 × 4-11 μm, thin walled. Cheilocystidia not differentiated.

Hyphae mostly inamyloid, without clamp connections. Hymenophoral trama of the *Boletus*-subtype.

Cortical layers: Epicutis of pileus a trichodermium (not palisadic) of mostly strongly interwoven hyphae which often form a pseudocutis in the upper region, the elements all strictly elongated between septa, cylindrical and 3-9 μm in diameter, with the yellowish brown or hyaline inamyloid walls 0.2-1 μm thick, some with internal fine granulosity which is hyaline to yellow brown, sometimes with resinous concolorous (yellowish brown) incrustation, the granular contents (where present) pseudoamyloid, and the whole layer at low magnification appearing deep rust brown; the terminal cells mostly 13-35 × (3)4-7.5 μm, in 20626 the incrustation where present, colorless in KOH and the cells vaguely cystidioid, cylindric like the other cells, or slightly swollen (subventricose, subfusoid, tapering to the obtuse tip, rarely narrowly clavate or subampullaceous in KOH). Covering of the stipe consisting of various types of dermatocystidia which basidiole-like, clavate, sterile e.gr. 19 × 7.2 μm, or fusoid e.gr. 26-38 × 7-11 μm, or ampullaceous e.gr. 49 × 7.5-12.5 μm, rarely tapering upwards from a broad base to an obtuse tip e.gr. 30 × 7.5 μm, without inside bodies or with internal granulation or guttulation or amorphous bodies, ± pseudoamyloid or inamyloid.

Chemical reaction: NH_4OH on pileus surface negative.

On the ground in montane oak woods (not *Q. oleoides)* fruiting in July.

Material studied: COSTA RICA: Cartago: 1.1 km E of "La Perla", Empalme, about 2000 m alt., July 1983, Gomez & Alfaro 20621 (F) type. — Same place and month, Gomez & Alfaro 20626 (F), see below.

This (the type) comes close to *Tylopilus pseudodecorus* (Snell, Mycologia 28: 22. 1936 as *Boletus*) Singer comb. nov. It. differs in less strongly reticulated stipe and its color, in slightly smaller spores, absence (or scarcity) of two-spored basidia on the stipe surface, smaller or poorly developed pseudocystidia in hymenium and on pileus, and mycorrhiza with different species of oak. Collection 20626 collected at the same time at the same place as the type, is just slightly different, mainly in the color of the pileus, and may be a form still closer to *T. costaricensis*. *T. pseudodecorus* is said to have unchanging context but in view of the fact that Snell considered this species identical with his version of *T. ferrugineus* may mean that the unchanging flesh should be taken "cum grano salis". We consider it possible that the collection by Gomez & Alfaro 20626 is conspecific with *T. ammiratii* Thiers, Calif. Mushrooms, Bol. (1975, p 221) in which case the latter might have priority over *T. costaricensis* which would be a variety. However, we find it disconcerting that the original description by Thiers states that the stipe overing is merely interwoven without (dermato-)basidia and caulocystidia. Furthermore, the hymenial cystidia seem to be longer. Our fungus (20626) has very variable but numerous and striking basidiole-like bodies on the stipe aside from the dermatocystidia and the hymenial cystidia are 24-42 × 7.5-11 μm; it occurs under Costa Rican life oaks. For these reasons we do not believe that our species is merely a variety of the Californian species.

16. **Tylopilus subniger** spec. nov. (Pl. 9, fig. 29; Pl.11, fig. 38-39; Pl. 12, fig. 40-41; Pl. 13, fig. 40 A, B, C, D).

Pileus at first almost black, later violet brown, dry and subtomentose, under a lens minutely scurfy or subpunctulate, at first velutinous, later very slightly rimose at times, not areolate but smooth, convex, eventually flat or concave, with a slight central depression, obtuse 40-118 mm broad.

Hymenophore tubular; tubes white, then more greyish or brownish, reaching 17 mm long, depressed around the stipe; pores white or creamy, soon more greyish, sordid or brownish when bruised, 0.5-1.2 mm wide, occasionally lamellarly extended at the stipe. Spore print not obtained.

Stipe fuligineous or dark chocolate because of a reticulation of this color on paler, below ± pale grayish ground, rarely staining yellow at the base, the reticulation reaching down to the lower third of the stipe, more rarely visible only in the upper two thirds, the meshes slightly elongated vertically, solid, tending to develop cavities, broader in upper than in lower half, generally with pseudorrhiza or (appendage) 60-165 × 15-31 (apex), 20-29 (base) mm. Basal mycelium pallid, eventually yellowish white. Veil none.

Context of pileus and upper part of stipe white, unchanging except in the zone just above the hymenophore where it becomes pale pinkish gray when bruised, or else becoming pinkish red when bruised, in the base tending to become yellow when bruised, or blackening, but also often unchanging. Tast mild. Odor weak, agreeable, in dried material with a slight cumarine smell.

Spores 10-15.5 × 4-6.5 μm, mostly 11-14 × 4-5.5 μm, rather versiform in younger specimens, later mostly boletoid, with or without a suprahilar depression or applanation, inamyloid, fewer pseudoamyloid, in KOH pale brownish melleous to subhyaline, smooth, without germ pore; Q = 2.1-3.5, mostly 2.3-3, average about 2.5.

Hymenium: Basidia 18-28 × 8.5-10 μm, 4-spored. Cystidia both in tubes and at pores 21-58 × 5.5-17 μm, ampullaceous, the apex thin (1-19 × 2.2-8 μm, more rarely cylindrical or clavate to fusoid, rarely somewhat subcapitate (capitulum to 3 μm diam.), most (pseudocystidia) with in Melzer chestnut contents, a minority (cystidia proper) hyaline granular and not pseudoamyloid.

Hyphae inamyloid, without clamp connections. Hymenophoral trama bilateral of the *Boletus*-type.

Cortical layers: Epicutis of pileus a trichodermium, not palisadic, consisting of interwoven, later rather loosely arranged (but not gelatinized) hyphal chains which are not incrusted, with brownish intracellular pigment and subhyaline walls (KOH), the terminal cells mostly cystidiform and ventricosely or claviformly widened, internally pseudoamyloid (spadiceous to porphyry brown in Melzer), but also (often the majority) not or not much changing in the Melzer, (27)20-55 × 7.5-13.5 μm, the lower cells of a chain mostly 4.5-11 μm broad and up to 65 μm long, not pseudoamyloid. Cortical layer of the stipe reticulum hymeniform consisting of ventricose to clavate cells 20-37 × 6.5-11 μm, dermatocystidia 23-42 × 4-13 μm, much like the terminal cells of the epicutis but often ampullaceous; dermatobasidia and basidioles scattered, 19-40 × 7-8 μm, 4-spored or 1-2-3-4-spored. All these elements frequently brownish inside (KOH).

Chemical color reactions: KOH on pileus orange. — NH_4OH on pileus negative.

In mixed frondose woods *(Quercus* and *Liquidambar,* or *Quercus* and

Magnolia) and in pure *Quercus* stands, in the montane zone in Mexico to 1700 m, in Costa Rica to 2200 m alt. Fruiting in June and July.

Material studied: MEXICO: Veracruz: near Banderilla, Rancho La Pomarrosa, 16 VI 1984. Anell 119 (F, XAL). COSTA RICA: San José, a few kms E of La Perla, about 2000 m alt. June 1983, Gomez 20627 (F). El Jardín, 8 VI 1984, Singer B 14555A (F), type. Cartago: La Chonta 17 VI 1983, Singer B 14387 (F).

According to García material from the Mun. de Acajete, leg. Bandala-Muñoz 1397, det. García is also conspecific with *T. subniger* which occurs in Veracruz in a zone between 700 and 1700 m, certainly associated with *Quercus* (as it is in Costa Rica).

This species may be related to *Boletus nigerrimus* Heim, Rev. Myc. 28: 281, 1963 but lacks olive or yellow colors on pileus and stipe; also the colors of the context are differently described. The generic position of *T. nigerrimus* is not altogether clear from the diagnosis.

17. **Tylopilus vinaceogriseus** spec. nov. (Pl. 10, fig. 33-37; Pl. 22, fig. 31; Pl. 23, fig. 00).

Pileus violet gray to vinaceous brown, eventually fuligineous or sepia, when dried 8 J 12 (M&P), between "Cochin" and "auburn" or between "Conga" and "auburn", dry, subglabrous, but under a lens finely and densely scurfy-granulose, with narrowly incurved margin, convex, obtuse, 60-110 mm broad.

Hymenophore tubular, dirty pallid; tubes about 6 mm long, ± depressed around the stipe; pores concolorous with tubes, becoming porphyry brown when touched or injured, less than 1 mm wide, some radially elongated in the zone adjacent to the stipe and often decurrent on it passing over into the network of the stipe.

Stipe in upper two thirds pallid, with a pallid, soon brown reticulation, in lower third without reticulation, in dried material when mature "trotteur tan" to (in places) "cattail" or "tiffin" glabrous, but in age coarsely innately fibrous, ventricose, ± "appendiculate" at base, 90-95 × 30-32 mm; basal mycelium dirty white. Context white turning distinctly pink on injury, in worm holes black; taste mild; odor none.

Spores (8)9-11.5(13.2) × 3.3-5 μm, mostly 3.5-4.2 μm broad, fusoid, more rarely cylindrical, even elongate-reniform or constricted, smooth, often with a elongated oil drop, melleous-subhyaline; inamyloid; those produced by dermatobasidia (10)10.5-14.5 × 4-5 μm, all with or without suprahilar applanation, Q = (2)2.7(3.3).

Hymenium: Basidia 18-25(38) × 5.5-8(9) μm, (2-) 4-spored. Pseudocystidia at the pores 20-37(60) × 7.5-9.5(12) μm, in tube interior 36-49 × 7.5-13 μm, with granular or globulose internal bodies or without them but all pseudoamyloid or at least orange brown to golden inside in Melzer, ampullaceous, rarely cylindric or fusoid. Cystidia without contents and inamyloid as well as true cheilocystidia not differentiated.

Hyphae without clamp connections, mostly hyaline, inamyloid; hymenophoral trama bilateral of the *Boletus*-type, mediostratum yellowish, lateral stratum hyaline, loosely arranged.

Cortical layers: Epicutis — a trichodermium (not a palisade), in some places later depressed to a "pseudocutis", terminal cells cylindrical or cystidiform, with or without internal granulosity or oil droplets, internally pseudoamyloid in many or most (deep rusty brown in Melzer), but always many which are only slightly pseudoamyloid, clavate and 28-31 × 6-10 μm or like the hymenial pseudocystidia, mostly 37-58 × 6-7.6 μm, the weakly pseudoamyloid ones often only 3-6 μm broad, lower cells 3-6.5(9) μm broad, some with a very fine incrustation hardly evident in KOH but pseudoamyloid; all these elements thin-walled, occasionally 0.5-1 μm thick walls observed, these subhyaline and inamyloid. Covering of stipe-network hymeniform, elements 6.5-27 × 2.5-7 μm, mostly 18-26 × 4.5-6.5, inamyloid; among them, some dermatopseudocystidia with hyaline wall and pseudoamyloid contents, the wall is somewhat thickened, dermatobasidia at apex smaller than those of the lower portion of the stipe, more like those of the hymenium, e.gr. 17 × 4.5 μm; in the non-reticulated lower portion of the stipe with the pavement cells e.gr. 16 × 8 μm, versiform, inamyloid, among them many dermatopseudocystidia, these deep rustbrown in Melzer, only very sparse dermatocystidia without pseudoamyloid contents.

Chemical color reactions. KOH on pileus brillant black; on stipe deep reddish brown; on tubes yellow-orange; on context: negative. — NH_4OH on pileus caesious, on context weakly gray; on stipe and tubes: negative — $FeSO_4$ on stipe dark gray-blue, on tubes caesious; on pileus and context negative.

On the soil under various species of *Quercus* (not *Q. oleoides)* in the montane zone, in *Quercus-Magnolia* forests. Fruiting early (June, July).

Material studied: MEXICO: Mexico: Mun. Tejupilico, Nanchititla, 3 VII 1988, R. Nava 177, det. García (F, ENCB). COSTA RICA: Cartago: La Chonta, 17 VI 1983, Singer B 14360 (F), type.

This species might be identical with *Tylopilus ferrugineus* subspec. *vinaceogriseus* Snell, Dick & Hesler but the latter was published without a macroscopical description. Thus, one may only judge by Snell and Dick's description of *T. ferrugineus.* This description is extremely close to our species if the color of sp. *vinaceogriseus* is taken into consideration. We believe however that at least the Costarican species is specifically different from *T. ferrugineus* (Frost) Sing. and therefore describe it as a new species rather than a new status of ssp. *vinaceogriseus.*

T. ammiratii Thiers, possibly related, is discussed under no. 15 (*T. costaricensis*).

18. **Tylopilus ferrugineus** (Frost) Sing. Am. Midl. Nat. 37: 106. 1947. (Pl. 14, fig. 42, 42A, 43, 43A, 44, 45).

Boletus ferrugineus Frost, Bull. Buff. Soc. Nat. Sc. 2: 104. 1874.

Pileus dark brown, then reddish brown to neutral brown (''sayal brown'', ''Verona brown'', snuff brown''), or thus colored from the beginning, not much changing color in dried specimens, not viscid, sometimes with a yellowish or orange bloom towards the margin, glabrescent there, entirely minutely tomentose, but often glabrescent, in young specimens sometimes very finely flocculose-tomentose, in age sometimes rimulose, pulvinate, then convex and often somewhat irregular, the margin sometimes very slightly projecting (but inconstantly so), 45-130 mm broad.

Hymenophore tubular; tubes whitish, then flesh color or cream, brown when bruised, 7-15 mm long, rarely longer, depressed around the stipe; pores concolorous with the stipes but sometimes becoming brownish to brown dotted, brownish ochraceous to brown when touched, at first roundish, later angular, 0.5-1 mm wide, isodiametric or subisodiametric. Spore print between ''isabella color'' and ''wood brown''.

Stipe whitish to "cream buff", but later subconcolorous with the pileus or somewhat lighter colored from the lower part upwards, with a reticulation which descends from the apex to one quarter to one third of the length of the stipe, more rarely to two thirds or not appearing at all, this reticulation white to avellaneous, or eventually brown where touched, the meshes longitudinally elongate, below the reticulation minutely punctate-pustulate to flocculose on paler ground, solid, versiform tapering upwards or downwards or subcylindric, central, rarely somewhat eccentric, 35-90 × 12-33 mm, rarely still larger. Veil none. Basal mycelium white.

Context white, becoming pinkish when young and fresh carpophores are bruised, or slowly changing to pinkish-avellaneous or ocher, eventually reaching brown color in wounds, at first very firm, but sooner or later becoming soft. Taste mostly mild, rarely slightly acrid. Odor none or rarely slightly farinaceous.

Spores (8)8.5-12.5(14) × 3-4.5(5) μm, most frequently 9.5-12.5 × 3.3-4μm, ellipsoid-subfusoid, some obclavate with tapering apex, with rather thin walls, inamyloid, with a slight or no suprahilar depression, subhyaline or slightly or slightly yellowish in KOH, few slowly subamyloid without germ pore and without truncation at the apex.

Hymenium: Basidia (18)22-29 ×7-10 μm, 4-spored. Pseudocystidia 24-72 × 7-19 μm, some only to 43 × 15 μm, clavate to fusoid or ampullaceous, rounded-obtuse or subacute, hyaline with subhyaline or yellowish, sometimes somewhat orangy granular contents which are pseudoamyloid (in some cases in old or broken cystidia, the contents material is seen in the medium or incrusting the walls, these thin; some cystidia are either granular inside or pseudoamyloid, but these are a minority. Cheilocystidia not well differentiated but like pseudocystidia, with a tendency to eventually browning walls and smaller size.

Hyphae without clamp connections, inamyloid. Hymenophoral trama bilateral of the *Boletus*-type, the mediostratum more pigmented than the lateral stratum. Oleiferous hyphae moderately numerous in the hymenophoral trama, 4-7.7 μm broad, sometimes connected with pseudocystidia.

Epicutis of pileus a trichodermium of interwoven hyphae, not forming a palisade, but sometimes all depressed to form a cutis-like surface in old specimens, terminal cells cylindric or cystidioid and then versiform, often utriform or subampullaceous, even clavate, often with short attenuate apex, and 26-72 × 4-9 or 4.5-13.5(22) μm, the lower cells mostly cylindric and 6-9 μm broad (a few cells short or swollen), with granular yellowish contents in most, this brownish orange to orange-brown in the Melzer. Covering of the stipe a subhymeniform layer consisting of dermatocystidia, either ampullaceous and much like the cystidioid end cells of the epicutis (so mostly in the reticulated portion), or rather short, ventricose-mucronate or clavate (so in the non-reticulate lower part), rarely as long as in the reticulate part.

Chemical reactions: KOH on pileus chestnut brown; on context yellowish to ochraceous; on pores as on context.

In frondose forests on the ground, in Mexico under *Quercus rysophylla,* also found in northeastern USA under *Fagus* and (in Japan) under *Castanopsis.*

Material studied: MEXICO: Tamaulipas, El Medroño, 1400 m alt, 24 IX 1985, García 5439 (F, ITCV). Also no 4813 ( ) - Nuevo León, Puerto Genovevo, 24 VII 1980, García 305 (F, ITCV).

Extralimital material was studied from New York, Massachusetts, North Florida and Georgia as well as from Kyoto, Japan (Singer A 4024, F).

T. ferrugineus is a well established species, the lectotype (leg. Frost, from Vermont USA) redescribed by Halling (1983).

A very similar species was described by Peck, as *Boletus indecisus* Pk *(Porphyrellus,* Gilbert, *Tylopilus,* Murr.) which differs from *T. ferrugineus* merely by narrower epicutis-hyphae (2.5-5 μm according to Wolfe 1981) and, in a average, larger spores. According to Snell (1970) the color of the pileus is slightly different, and the context is less firm, but these appear to be rather variable characters.

19. **Tylopilus indecisus** (Peck) Murr., Mycologia 1: 15. 1909.

Syn.:

Boletus indecisus Peck, Ann. Rep. N.Y. State Mus. 41: 70. 1988.
Porphyrellus indecisus (Peck) Gilbert, Bol., p. 99, 1931.

Pileus light or light cinnamon brown (''kis kilim'' when dried, Maerz & Paul), often finely rivulose (pallid in the fissures), smooth, subglabrous, convex or partly applanate, 110 mm broad.

Hymenophore tubular, depressed around stipe, tubes pink, pores concolorous with tubes and like these becoming reddish brown when bruised, almost round, small. Spore print not obtained but apparently pinkish.

Stipe at apex slightly reticulate, minutely pustulate below the reticulum, the ornamentation light brown on whitish or pale brown ground, subcylindric, 110 × 35 mm; veil none; basal mycelium white.

Context white or whitish, slightly ochrascent when bruised; odor agreeable; taste mild.

Spores (9.5)10.5-15.5(18.5) × (3)3.5-4.5 μm, cylindric to fusoid, rarely oblong (Q = 2.5-4.3), pale melleous yellowish (KOH), smooth, inamyloid to ± pseudoamyloid, not truncate and without germ pore, wall to 0.4 μm thick, in an average slightly larger and slightly narrower than those of the preceding species.

Hymenium: Basidia 19-33 × 5-9.5 μm, (2) 4-spored.

Pseudocystidia rather numerous especially near pores, 40-91 × 10-12(15) μm, fusoid to fusoid-subampullaceous, thin-walled, with in KOH melleous guttulate, in Melzer rusty-chestnut contents, in Melzer often appearing internally subreticulate or with inamyloid parts of the interior, few totally inamyloid (cystidia).

Hyphae: inamyloid, without clamp connections in the bilateral *(Boletus*-subtupe) hymenophoral trama.

Covering layers: Epicutis of the pileus a trichodermium, the outermost layer rather loosely arranged but not gelatinized, hyphous, the hyphae with chestnut brown intraparietal pigment, 2-4.5(6) μm thick, but locally (apically or rarely in the middle) slightly inflated to 7 (8) μm but the majority filamentous, smooth in KOH, with rounded tip, the trichodermium 200-250 μm deep; hypodermium not differentiated. Covering of the stipe with some dermatopseudocystidia like the hymenial pseudocystidia and a few basidia like the hymenial basidia.

Under oak in mixed forests. In Mexico fruiting in June, rather rare.

Material studied: MEXICO: Nuevo León: Mun. Santiago, 3 km along road Potrero Redondo-La Trinidad at 1500 m alt., 9 VI 1979, García 22, (ITCV, F).

This species, already indicated by García for Mexico in 1981, is less common than *T. ferrugineus* south of the border with U.S.A. It is very similar to the latter, and differs mainly by the narrower elements of the epicutis and the somewhat larger and narrower spores, perhaps also by the colors and a slighter discoloration of the context. The description of *T. indecisus* has been completed, but is based exclusively on Mexican material, yet it shows perfect identity with the data published on Peck's type by Wolfe (1981). One might be tempted to reduce this species to a variety of *T. ferrugineus,* but this would not serve any useful purpose at present.

García has observed giant spores up to 25 μm in length.

Illustrations of the Mexican material were published in Bol. Soc. Mex. Mic. 15: 184, fig. 120-122 and 192, fig. 153 (in color) by García & Castillo, and this was found to be in good agreement with Snell & Dick (1970), pl. 86, fig. 1 and pl. 61 (in color).

20. **Tylopilus hondurensis** Sing. & Ivory, Beih. Nov. Hedw. 77: 104, plate 12 (p. 146). 1985.

Pileus pale hazel (about p. 13 17 Maerz & Paul according to photo), with somewhat paler marginal zone, dry, minutely cracked, dried appearing rivulose, under a lens with subreticulate appressed minute flocculi which are dark on pallid ground, convex, about 90 mm broad.

Hymenophore tubular, tubes buffish, about 10 mm long, adnexed; pores pale luteous, apparently unchanging, subisodiametric, somewhat angular, small. Spore print drab when dry.

Stipe white with light hazel reticulation and striations less prominent than in *T. felleus,* with a vinaceous patch near base, subequal with slightly narrowed apex and base, the lower half or third smooth, glabrous but when dried under a lens minutely dotted sepia or gray-brown on pallid ground, solid, about 100 × 15 mm. Basal mycelium white.

Context white, unchanging, firm-fleshy in the pileus. Taste slightly bitter. Odor pungent.

Spores 12-13.5(14.3) × 3.8-5(5.2) μm, fusoid, smooth, with suprahilar depression, almost or quite inamyloid when seen singly but in superposition rather distinctly though weakly pseudoamyloid, cyanophilic.

Hymenium: Basidia 21-21.5 × 8.7-10 μm; pseudocystidia 21-28 × 6.5-10 μm, clavate, with guttulate or amorphous contents appearing pseudoamyloid (red-brown) in Melzer's reagent.

Hyphae of the pileus trama hyaline, inamyloid, without clamp connections, hymenophoral trama bilateral of *Boletus*-type.

Cortical layers: Epicutis of the pileus, not or scarcely differentiated from the hypodermium, formed by a trichodermium consisting of elongated elements with a spadicous vacuolar pigment (KOH), the trichodermium about 100 μm deep, the terminal cells mostly cystidioid, these dermatocystia 33-73 × 5.3-10 μm, either short cylindric and rather small or fusoid and then often long-pedicellate and large, or else clavate to subclavate and medium sized. Covering layer of the stipe (on reticulating veins which are composed of longitudinally repent filamentous hyphae) hymeniform, consisting of dermatobasidia and dermatocystidia, the dermatobasidia e.gr. 10.7 × 3.3 μm, 4-spored, producing spores often somewhat smaller than those produced in the hymenophore; dermatocystidia either like those of the tube-hymenium or else ampullaceous, e.gr. 21-21.5 × 6 μm with the neck only 1.3-1.5 μm across, with some melleous-succineous granular or guttulate, somewhat pseudoamyloid internal bodies.

Solitary under *Pinus oocarpa,* with grassey ground cover.

Material studied: HONDURAS: Siguatepeque, 12 X 1976, Ivory S/249 (F), typus.

This species appears to be close to *T. griseopurpureus* Corner and *T. tabacinus* var. *amarus* Sing. The former is a palaeotropical species and differing from *T. hondurensis* in colors, cespitose habit, differently shaped and larger cystidia, and the pigment is described as what we would call intraparietal rather than vacuolar. As for *T. tabacinus* var. *amarus,* a Floridian form, the darkening of the context by autoxidation refers it to the group known as sect. Ferruginei; likewise, the pigments of the epicutis and the habitat are quite different.

21. **Tylopilus brachypus** spec. nov. (Pl. 15, fig. 46-51).

Pileus pale brown or stramineous brown, frequently areolate (appearing almost appressedly squamulose), often in part subglabrous (not areolate) and may be initially with entire cuticle, paler between areolae, not rugose or sulcate, with not or scarcely projecting sterile margin, convex, obtuse, (dried) 70-95 mm broad.

Hymenophore tubular, tubes 5-6 mm long when dried, very short at margin, pale brownish with a pink shade when mature, pores isodiametric and 0.7-1.4 mm wide near margin (dried), but toward stipe radially elongated and at stipe sublamellar, there the hymenophore adnexed to adnate and not or scarcely depressed. Spore print not obtained.

Stipe above distinctly reticulate and yellowish, with the meshes longitudinally elongate, reaching down 15-30 mm, and from there replaced by a non-reticulate zone which, instead is light to dark fuscous, in dried material gray dotted on paler ground down to the very base, dry, solid, strongly tapering downwards and remarkably short, 35-66 × 21-25 mm (dried); veil none; basal mycelium white or whitish. — Context whitish, when bruised pink, thick-fleshy in the pileus.

Spores 11-14.5 × 3.5-5.5(6) μm, mostly fusoid, fewer wedge shaped with broadened base, with very scanty giant spores (to 17.5 × 7 μm), pale melleous, smooth, inamyloid, others pseudoamyloid.

Hymenium: Basidia 21-29 × 8-10.8 μm, 4-spored. Pseudocystidia on pores and in tubes moderately numerous, versiform, but mostly clavate to ampullaceous and 22-40(60) × 7-11.5 μm, chestnut inside in Melzer medium; some few without pseudoamyloid contents (cystidia). Cheilocystidia not differentiated.

Hyphae inamyloid, without clamp connections in the carpophore, but in the hymenophoral trama (of the *Boletus*-type) many false clamps observed, and in the stipe some few septa are found to be regularly clamped.

Covering layers: Epicutis of the pileus a rather thin layer, often intermittent (where areolate), a trichodermium of loosely arranged hyphae 2-5(7.5) μm thick, usually filamentous and strictly elongate, rarely the terminal cell claviform, 6.5-7.8 μm diameter, not incrusted (KOH), many cells pseudoamyloid, very few almost spheroid in KOH fuscous or fuscidulous, the pigment often colloidly condensed. Hypodermium a less colored or hyaline cutis-like stratum, no gelatinisation demonstrable in the cuticle. Covering of the stipe showing the longitudinal hyphae ascending to a layer of dermatocystidia which are mostly clavate, some broadly ventricose and mucronate or subcylindric, all thin-walled and hyaline or subhyaline (KOH) and forming a hymeniform or subhymeniform layer, in the reticulate part and in the upper zone of the dotted part of the stipe with dermatobasidia, forming slightly smaller spores than the hymenial basidia. The dermatocystidia are often rising from small isodiametric basal cells of 5-10 μm diameter (one to three). The pigment not visible in KOH mounts, but the contents of the dermatocystidia chestnut or at least mostly distinctly pseudoamyloid, few inamyloid.

In a montane forest of *Pinus* and *Quercus*.

Material studied: MEXICO: Durango: Between Potrero Mesa Larga and Arroyo Presa Los Altares in the Reserva de la Biosfera de la Michilia 2500 m. alt. 4 IX 1983, Valenzuela 2470 (F, type, also ENCB).

This species is obviously closely related to *T. hondurensis* Sing. & Ivory (no. 20). Unfortunately, we have no data on the mycorrhizal host (*Pinus* or *Quercus*), no spore print and no indication on the taste of fresh specimens. Nevertheless, we believe that the shortness and strong tapering downward of the stipe, the deeper epicutis with cystidioid terminal cells, and the presence of ampullaceous dermatocystidia with narrow apex on the stipe are sufficient ground to separate these two species. Only if *T. brachypus* should be shown to form ectomycorrhiza with pine, has the same spore print color and taste as *T. hondurensis,* and ocurs also on lower altitudes it might be possible to consider the Mexican species as conspecific with the Honduran one, allowing strong variations in stipe size and shape.

Supplement

Species without pseudocystidia

Species fully coinciding with the characteristics of the genus *Tylopilus* but not showing any pseudocystidia in the hymenium or the covering layers are temporarily inserted here. It is quite possible that such species not referable to either *Austroboletus* (because of smooth spores) or *Fistulinella* (because of habit and lack of gelatinascant layers of the epicutis) or *Veloporphyrellus* (because of the absence of a veil and presence of abundant pigment) belong in an additional section of *Tylopilus* and that the presence of pseudocystidia has been overemphasized in the past. We describe here (p. 72) a pseudocystidiate *Fistulinella,* and the presence of pseudocystitia is also known in some *Austroboletus.* We do not formally describe a new section for the pseudocystidium-free *Tylopilus* because our knowledge of *T. corneri* is incomplete and a further collection belonging in the same group comes from Borneo and has no macroscopical descriptive data.

22. **Tylopilus corneri** Sing. spec. nov.

Pileus with chocolate brown areoles on buff ground, dry, convex with eventually applanate center, about 77 mm broad.

Hymenophore greyish brown, tubulose with rather small pores which are somewhat radially elongated at the margin of the pileus.

Stipe pearl gray, smooth except for an apical region 6-7 mm wide where it is coarsely but shallowly ribbed-subreticulate at the extreme apex, 52 × 35 mm; veil none.

Context whitish, becoming red when bruised, eventually darker. Taste mild in dried material; odor Ganoderma-like but weak.

Spores 9.5-15 × 4-7 μm, mostly 10.5-14 × 4.5-5.5 μm, more or less fusoid or oblong, rarely ellipsoid, with suprahilar depression but sometimes with applanation, in light microscope quite smooth, usually without germ pore, rarely with a very slight discontinuity at apex, wall 0.2-0.4 μm thick and very pale melleous, the interior still paler or pale (KOH) and palest dirty livid in Melzer when mature, the episporium dark, not truncate.

Hymenium: Basidia 26-39 × 8.5-15 μm, 4-spored. Cystidia 40-42 × 9.5-16 μm, ventricose or ventricose-clavate in tube interior, some shaped like cheilocystidia at pores, thin-walled, optically "empty". Cheilocystidia 20-28(40) × 4-7.5 μm, fusoid and acute or obtuse at the tip, some slightly incrusted at apex, without visible contents and inamyloid.

Hyphae inamyloid, without clamp connections.

Cortical layers: Epicutis a fractured trichodermial palisade, which is moderately deep and consisting of elongatd, fusoid to cylindrical or narrowly ventricose parallel hyphal cells which have thin-walled, obtuse terminal cells with hyaline wall and no incrustation, with dissolved, rather pale porphyry or chocolate brown intracellular pigment only, not granular and inamyloid in the Melzer. Hypodermium poorly differentiated. Covering of stipe with a now disrupted palisade of somewhat broadened cells.

Under oaks on the ground.

Material studied: COSTA RICA: San José: 6 km west of Empalme, between 1800 and 2000 m 2 VII 1983. Gomez & Alfaro 20582 (F), type.

The material is in rather poor condition, but the microscopical analysis shows it to be quite different from any Costarican *Tylopilus.* It keys out with *T. alboater* Schwein. sensu Corner non Schwein. (which should presumably be called *Boletus pachycephalus* Mass.) from which it differs in less distinctly blackened, broader bulbous stipe, different cheilocystidia, larger cystidia, intercellular pigment and slightly broader spores. It also differs from *B. pachycephalus* Mass. as originally described by Massee in not unchanging context. Ours is apparently a new species which is closely allied to a Borneo collection (see below). It seems to be close (but not identical) to *Boletus subunicolor* Dick & Snell but has larger spores and broader stipe, see Wolfe (1978).

Extralimital Species

A specimen determined *Boletus phaeocephalus* Pat. & Baker by Corner and as such cited (Corner 1972) does not belong to *Pulveroboletus phaeocephalus* (Pat. & Baker) Sing. but rather to *Tylopilus.* The macroscopical characters are probably more or less as given by Corner p. 129. l.c.

The hymenophoral trama is now cinnamon-pallid, the pores subisodiametric but somewhat radially elongated (0.3-0.6 × 0.1-0.3 mm dried) especially toward the margin, and the surface of the pileus appears brown, yellow brown, glabrous and smooth.

Spores 9-15 × (3.5)4-5.7 m, fusoid with suprahilar depression or applanation, with attenuate-subobtuse or more rarely subacute apex, melleous-hyaline, very pale in KOH, many of them with strongly pseudoamyloid contents, young spores inamyloid, smooth, with one elongated or 2-3 small round oil drops, not truncate at apex and without germ pore, the wall up to 0.7 µm thick.

Hymenium: Basidia 15-33 × 9-13 µm, 4-spored, occasionally some 2- or 3-spored ones intermixed. Cystidia 21-32 × 6.5-9.5 µm, fusoid with subacute or obtusate tip, or ventricose-mucronate or ampullaceous-subcapitate, moderately numerous, hyaline, thin-walled, without visible contents and inamyloid. No cheilocystidia or pseudocystidia differentiated.

Hyphae without clamp connections, hyaline, inamyloid. Hymenophoral trama with non-gelatinized melleous mediostratum and moderately gelatinized recurved hyaline hyphae 1-5 µm broad in the lateral stratum, thus of the *Boletus*-subtype. Hyphae of the pileus filamentous, hyaline 2-16 µm broad, versiform, interwoven, oleiferous hyphae present.

Cortical layer of the pileus: Epicutis a trichodermial palisade, the terminal cells cystidiform, of two types (a) 20-23 × 4-8.5 µm, like the hymenial cystidia — (b) 20-25.5 × 8-14 µm, clavate, sometimes with a sublateral sterigmatoid appendage, both types hyaline, inamyloid, thin-walled and often rising from a short basal cell intercalated between epicutis and hypodermium. Hypodermium a deep trichodermium, consisting of elongated yellow cells which are interwoven, without intraparietal pigment and without incrustation and thin walls.

Material studied: BORNEO: Kinabalu, 1300-1600 m alt. 14 III 1964, Corner (K).

This species is regarded as near *Tylopilus corneri* Sing. (no. 22). It differs from it in shorter cystidia and minutely cracked pileus as well as in less bulbous stipe. Cheilocystidia did not appear to be differentiated but pleurocystidia slightly longer than indicated above may occur.

The species is probably undescribed. Without more data on the fresh fungus we do not wish to introduce a new species.

III. AUSTROBOLETUS (Corner) Wolfe

Bibl. Mycol. 68: 64. 1979.

Syn.: *Boletus* subg. *Austroboletus* Corner, Bol. Mal. p. 76. 1972.
Porphyrellus sect. *Tristes* Sing. Farlowia 2: 115. 1945.
Porphyrellus sect. *Graciles* Sing. l.c. p. 119.

The characters of this genus are evident from the key and are discussed here on p. 11.

Key to the sections

1. Surface of stipe alveolate, reticulate-winged, often lacerate and scale-like. . . . . sect. AUSTROBOLETUS
1. Surface of stipe striate, ridged longitudinally, sometimes subreticulated to low-reticulated. . . . . sect. GRACILES

Section AUSTROBOLETUS. The type of this section (and the genus) is *B. dictyotus* (Boedijn) Corner.

Description see key above.

1. **Austroboletus neotropicalis** spec. nov. (Pl. 17, fig. 52).

Pileus olive, olive-yellow, later yellowish brown to leather brown, at first somewhat hispid-felty, later rimose-tomentose with whitish to yellowish tones between the eventually frequently appearing subsquamose areolae, with at first almost veil-like projecting sterile margin, dry, convex, obtuse, 17-50 mm broad.

Hymenophore tubular; tubes whitish to livid, becoming pink or cream up to 13 mm long; pores pink at maturity and when dried, 0.5-0.8 mm wide, depressed and slightly short-lamellate at stipe, on bruising somewhat livid violet to violet brown, ± angular. Spore print violet-brown to almost chocolate.

Stipe white or whitish, often somewhat yellowish, especially at apex, with an alveolate and often somewhat lacerate reticulation, sometimes with olive colored outer margins which turn brown when bruised, the ornamentation usually thinner at th apex than otherwise, slightly to distinctly gradually tapering upwards, generally solid, 60-160 × (3)-5 mm; veil not observed; basal mycelium white.

Context white, unchanging or very slightly yellowing; taste bitterish to strongly bitter; odor "fungoid" or none.

Spores (12)13-16 × (5.5-6.5)6.8-8.5 μm, mostly 13.5-14.5 × 7-7.7 μm, fusoid with subacute apex, with short and straight or curved ridges which are lower towards apex and base, but the ornamentation as a whole becomes deeper as the spores mature, with wall and ornamentation together measuring 1.2-1.5 μm, ornamentation of type XI, brown on pale yellowish ground, pseudoamyloid, Q = 1.73-2.00.

Hymenium: Basidia 22-37 × (.5)10-14.5 μm, (2)-4-spored. Cystidia rather scattered (cheilocystidia not differentiated), 30-63 × 8-10(20) μm, hyaline and thin-walled, occasionally with slightly thickened wall, rare to moderately numerous, fusoid, without discernible contents and wall as well as contents inamyloid.

Hyphae without clamp connections, inamyloid. Hymenophoral trama of the *Boletus*-type. Oleiferous hyphae about 6.5 μm broad.

Epicutis of pileus of long peg-like structures which are hyphal structures forming a hispid covering or intermittent trichodermium passing into strands of subparallel hyphae (1)2-8 μm thick and occasional cystidioid hyphal ends which are 40-80 × 4-10 μm inamyloid, not gelatinized, hyaline without visible contents, with the wall thin or up to 0.5 μm thick, some brownish incrusted by a very low incrustation. Covering of stipe hyphous, of filiform to very broad hyaline cells, not hymeniform.

Chemical reactions: KOH on pileus chestnut brown; on stipe violet brown, on context negative.

On ground under *Quercus,* often among fallen leaves, in Costa Rica under *Q. oleoides.*

Material studied: MEXICO: Tamaulipas, Rancho El Cielo, Municipio Gomez-Farias 4 VI 1989, 9 VII 1984 Garcia 6189 (F, ITCV). Also

Veracruz, Gardens of INIREB, García 4766 (ITCV). Municipo de Chiconquiaco, Ventura 8637 (ENCB). — COSTA RICA: Guanacaste, 7 km NW of Bagaces, 80 m alt., X 1982. Gomez 18708 (F), type.

The material was originally misdetermined by us as *A. subflavidus* (Murr.) Wolfe, but is obviously different (see Singer 1945 and Wolfe 1979). A similar form was determined *A. subvirens* (Hongo) Wolfe. All three species are indeed very closely related and *A. subvirens* ss. Halling (Mycotaxon 34: 94. 1989) is a fourth to enter the complex. According to the published data, they may be distinguished by the following key.

1. Pileus subviscid, large: 40-85 mm, more than half as broad as length of stipe; stipe yellowish with olive network; taste bitter. Eastern and Southeastern Asia.. .... *A. subvirens*

1. Pileus not viscid, and without any greenish or olive colors; or else diameter of pileus up to half as broad as stipe length; stipe white with yellowish apex and sometimes olive colored margins of raised reticulum. Taste bitter, bitterish or mild. Western hemisphere species.

   2. Pileus large, 47-107 mm, this and stipe without any greenish or olive colors. Florida species with *Quercus lauriforlia, Q. virginiana* and *Q. minima*....... .... *A. subflavidus*

   2. Pileus 15-50 mm broad, at least when fresh and young distinctly olive colored on pileus and often also on stipe. With other species of oak, not in Florida.

      3. Pileus 15 mm broad; stipe 30 × 3-4 mm, white, green above, ± subequal; spores 14.7-18 × 6.3-8.8 μm; taste mild. South America............... .... *A. subvirens* ss. Halling

      3. Pileus 17-50 mm broad; stipe 60-160 × (3)5-17(20) mm, white or yellowish, especially at apex, the margins of the reticulum sometimes olive colored, slightly to distinctly gradually tapering upwards. Spores (12)13-16 × (5.5)6.8-8.5, mostly 7-7.7 μm broad, i.e. slightly broader than above. Taste bitter.* Mexico to Costa Rica..................*A. neotropicalis*

*The taste was indicated as mild in specimens from Tamaulipas. It is difficult to know if this was in error or indicates some variation. The Costa Rican specimens were decidedly smaller (although within the range of *A. neotropicalis*) than the Mexican material. Again, we cannot judge whether this is an approximation toward the Colombian form, or a coincidence. At any rate, the Costa Rican specimens were strongly bitter.

Sect. GRACILES (Sing.) Wolfe, Biblioth. Mycol. 69: 69. 1979.

Porphyrellus sect. *Graciles* Sing. Farlowia 2: 119. 1945.
Tylopilus sect. *Graciles* (Sing.) Smith & Thiers, Bol. Mich. p. 94. 1971.

Characters, see key to the sections (p. 53).

Key to the species

1. A minority of the spores distinctly ornamented when seen in light microscopy; pileus sordid white to blackish gray. 2. *A. heterospermus*

1. Majority of spores distinctly ornamented even in light microscopy; pileus somehow brown (mostly chestnut brown or tawny, or finally paler, or more cinnamon). . . . . 3. *A. gracilis*

2. **Austroboletus heterospermus** (Heim & Perreau) Sing., Nova Hedwigia Beiheft 77: 135. 1983.

Syn.: *Porphyrellus heterospermus* Heim & Perreau. Bull. Soc. Myc. 80: 93. 1964.

Pileus sordid white to blackish-gray, smooth, opaque, with young recurved margin, strongly convex, regular in shape, reaching 130 mm in diam.

Hymenophore slightly convex, tubes 8 mm long in a young specimen, depressed but continuous around the apex of the stipe, whtish, pores small, dark maroon with a copper tinge, with sordid cream colored edges, ending near stipe in lamellate segments; spore print color not indicated.

Stipe concolorous with pileus but a little paler, longitudinally striate, central, rather robust, cylindric but slightly thickened at the base and somewhat curved 70 × 20 mm in a young specimen.

Context of pileus thick, white under the black cuticle becoming gray then pinkish when cut; white and fibrous-silky in the stipe. Odor none. Taste not very agreeable but not bitter.

Spores (12.7-)13.5-16(17.5) × (5.4-)5.8-6.2 μm, fusiform, long elliptic, or cylindric, with or without a supra-apicular, slight depression, with rounded more or less extended tip, some with median constriction, smooth, or more rarely ornamented distinctly verruculose or the wall showing darker spots like in the preceding species, or forming rather high (1-1.5 μm) tubercles, all ornamentations episporial, the wall 0.7-1 μm thick, with reddish-gray or greenish-gray endosporium and a thick reddis-yellow, towards the outside more black episporium, general coloration under the microscope light reddish-brown, finely granular or with numerous small droplets inside.

Hymenium: Basidia 50-55 × 13-14 μm, 4-spored, clavate; cheilocystidia 60-100 × 13-18 μm, hyaline, claviform or fusiform, at apex generally covered by granulations or particles but not muricate; pleurocystidia numerous, 70-80 × 12-14 ×, with granular contents, hyaline, thin-walled, fusiform-elongate, acuminate or claviform with one appendage.

Hyphae: Hymenophoral trama slightly bilateral, almost regular. Hyphae of the content without clamp connections, thin-walled, 6-7 μm thick, hyaline. Covering layer of pileus, a trichodermium formed by erect hyphae which are more or less interwoven 6-14 μm broad, with granular contents and thin wall, the terminal members elongated with rounded tips, clavate, attenuated or capitate.

Chemical characters: Phenol: blackish purple; $FeSO_4$: slightly green.

In *Abies religiosa* forest.

Material studied: None. Heim's description, as given above, shows clearly enough that this species belongs in *Austroboletus* as characterized by Corner, Wolfe, and recognized in the present paper. Since no type was available and the spore print color is unknown, the description should be completed by studies of the type (if available) or a topotype. The material came from MEXICO: D.F., La Venta, W of Mexico 13 VIII 1961, 3100 m alt.

Unfortunately, the alkali reaction with the context of *A. heterospermus* is unknown, but the fact that only part of the spores show any ornamentation under the light microscope reminds one of the observation — confirmed by Wolfe (1979, p. 30 and Moore & Grand 1970) — that partially ornamented spores can occasionally be observed among hundreds of smooth spores of species of the *P. prophyrosporus* group where a reddish alkali reaction is often observed (as in *Stro-*

bilomyces). These were the original reasons for linking *Porphyrellus* s.str. with *Austroboletus,* and arranging these genera in the Strobilomycetaceae. And to this day, it must be admitted that there is a transition between *Tylopilus-Porphyrellus-Austroboletus-Fistulinella* inasmuch as the tendency of the young pileus-trichodermium to gelatinize in *Austroboletus* is not a general character of the latter genus.

3. **Austroboletus gracilis** (Peck) Wolfe Bibl. Mycol. 69: 69. 1979.

Syn.: *Boletus gracilis* Peck, Ann. Rep. NY St. Mus. 24: 78. 1872.
Tylopilus gracilis (Peck) Henn. *in* Engl. & Prantl. Nat. Pfl. fam. 1[1**]: 190. 1897.
Porphyrellus gracilis (Peck) Sing., Farlowia 2: 121. 1945.

Pileus chestnut, orange brown or yellow brown ("antique gold" M. & P.), or cinnamon, velutinous to subtomentose, in age after rains slightly viscid, becoming rimulose to subareolate, often scrobiculate pulvinate, then ± convex, margin ± projecting (20)40-80(100) mm broad.

Hymenophore tubular, tubes white, later pink, 6-13 mm long, depressed and lamellarly extended around the stipe; pores concolorous with the tubes, 0.5-1 mm wide, more often 1 mm wide, angular, whole hymenophore unchanging when bruised. Spore print pinkish brown, reaching 6/7 D/E 6/7 Kornerup & Wanscher, according to Wolfe, dried after a while "Sudan brown" (Ridgway).

Stipe somewhat paler or concolorous with the pileus, slightly pruinose to granular-furfuraceous, with or without longitudinal ridges over part of or all the stipe and these ridges often very finely (lens) reticulated, solid, subequal or tapering upwards, often curved, 60-130 × 6-20 mm. Veil none. Basal mycelium white.

Context white, unchanging on unjury, soft in the pileus, sometimes cinnamon pink underneath the cuticle. Taste mild. Odor none.

Spores 10-17(17.5) × 5-7.7(9) µm, oblong to ovoid-oblong, melleous brown to melleous with slight or distinct suprahilar depression, without germ pore or apical truncation, some pseudoamyloid, punctate when seen from above because an ornamentation of type XI.

Hymenium: Basidia 24-30(36) × (7.5)9.8-11.8 µm, 4-spored or (2)-4-spored. Cystidia moderately numerous, 32-50(90) × 5-11 µm, fusoid to subampullaceous, hyaline, thin-walled, inamyloid. Cheilocystidia 20-50 × 6-12 μm, narrowly clavate or clavate-ventricose, otherwise like pleurocystidia.

Hyphae without clamp connections, inamyloid. Hymenophoral trama bilateral, of the *Boletus*-type.

Epicutis a slightly palisadic trichodermium, becoming eventually an ixotrichodermium, consisting of chains of inamyloid hyphae mostly 5-10 μm broad, some rather short (not spherocystoid) with the terminal cell often cystidioid and much like the cheilocystidia. Covering of the stipe with dermatocystidia much like the cheilocystidia, forming a hymeniform layer where the stipe is reticulated.

Chemical color reactions: KOH and NH_4OH on pileus: Paler. — $FeSO_4$ on context greenish.

On the ground in mixed woods of pine and other conifers, birch and oak, or in oak woods mixed with other frondose trees.

Material studied: MEXICO: Nuevo León: Mun. Zaragoza, Cerro del Viejo, 2900 m alt., 26 IX 1982, García 2593 (F). — Durango; road Durango-Mazatlán, km 122, 2650 m alt., 28 VII 1984, García 4134 (F, ITCV). — Veracruz: SW of Banderilla, Cerro La Martenica 27 VI 1984, Guzmán 24418 (F, ENCB). — Morelos, Parque Nacional Lagunas de Zampoala 23 IX 1970, Guzmán 8373 (F, ENCB). — Guerrero: Mun. Chichihualco, road El Carrizal. — Atoyac, km 4.5, 2600 m alt., 15 VII 1983, Perez Ramirez 426 (F, FCME). — Michoacán: road Morelia - Cd. Hidalgo, km 40, 21 VII 1983, García 3662 (F, ITCV). — Mil Cumbres, 9 VIII 1969, R. Singer M8993 (F). — Also: Many collections from U.S.A.

IV. PORPHYRELLUS Gilbert

Les Bolets p. 99. 1931.

Syn.: *Phaeosporus* Bat. Bolets p. 11. 1908 non Schröter (1888).
Boletus subgen. *Porphyrosporus* Smotlacha, Mon. Cesk. h. hrib. p. 31. 1911.
Tylopilus subgen. *Porphyrellus* (Gilb.) Smith & Thiers, Bol. mic. p. 94. 1971.

Description see Singer (1986) and key (p. 3).

We recognize *Porphyrellus* on the generic level. Aside from the morphological reasons enumerated by Singer (1968, p. 792), a biogeographical one was indicated by Singer (1988) which is corrobated by the data of the present monograph. The species of *Porphyrellus* do not appear among those making up the mycoflora of Costa Rica and are not known to occur south of the American area of *Pinus*. They form ectomycorrhiza with Pinaceae and Fagaceae only. Their origin is in the north temperate zone. Their biogeographical position in relation to *Tylopilus* is what the relation of *Gyrodon* is to *Phlebopus* and *Meiorganum*. (We tentatively exclude sect. *Novazelandianus* Wolfe*, sect. *Africanus* Wolfe and *T. amylosporus* Smith which should be restudied from the type locality, see Smith 1965**).

Key to the species occurring in Mexico

1. Stipe reticulated over some distance from the apex....... 4. P. ZARAGOZAE
1. Stipe reticulated only at the extreme apex as a continuation of the tubes, or not reticulated at all.
   2. Cystidia not pseudoamyloid (only dermatocystidia especially of the stipe slightly pseudoamyloid).
      3. Spores (10)10.8-13.3(15) × (4.8)5.2-6.3 μm. Mycorrhiza with oak....... .... 2. P. UMBROSUS

*Wolfe (1986a) in the diagnosis of the section writes (pseudocystidia ...) adsunt (p. 518). Obviously *absunt* is meant since his type studies state that they are absent.

**"differs dramatically from all other OZUs" (Wolfe 1986, p. 518).

3. Spores (9.5)10.5-16(19.5) × (4.5)5-7.2(9) µm. Mycorrhiza with conifers. . ... 1. P. PACIFICUS

2. (Pleuro)cystidia (pseudocystidia) immediately or slowly pseudoamyloid (a minority at times not).

4. Spores small (mostly not over 15.5 µm long); cells of epicutis often ventricose and up to 21 µm broad. Mainly in oak woods. .... 3. P. CYANEOTINCTUS

4. Spores at least to 19 µm long; epicutis elements 7-14 µm broad. In mixed forests. .... 5. P. PORPHYROSPORUS

1. **Porphyrellus pacificus** (Wolfe comb. nov. (Pl. 16, fig. 53-56).

Syn.: *Tylopilus pacificus* Wolfe Biblioteca Mycol. 69: 44. 1979.

Pileus brown to olive brown, "cocoa br" when dried, dry, finely fibrillose-tomentose, soon rimulose areolate, whitish between areolae, convex, 65-80 mm broad.

Hymenophore tubulose, tubes greyish brown to brown, about 11 mm long, changing to blue, greenish blue when bruised, but sometimes greenish towards the margin or partly olive brown, slightly depressed around the stipe; pores concolorous or somewhat green towards the margin and with an olivaceous brown, dry darker brown tinge, otherwise angular, 0.5 to about 1 mm in diameter. They turn greenish blue or green when touched. Spore print not obtained.

Stipe at first whitish brown, then greyish brown to olivaceous brown, smooth (not reticulated), solid, cylindrical, to 113 × 15 mm. Veil none.

Context pallid, bluing (?), at pileus margin turning violet or lilac, thick fleshy. Odor fruity; taste mild.

Spores (9.5)10.5-16 (19.5 according to Wolfe, type, 1979) × (4.5) 5-7.2(9) µm, ellipsoid or ovoid-ellipsoid to subfusoid or subcylindroid, often narrowed to subacute above, with suprahilar depression or applanation or without it, smooth without truncation or germ pore, olive yellow to brownish in KOH or NH_4OH, subamyloid, pseudoamyloid or inamyloid, wall ± 0.5 µm, more rarely to 1 µm thick.

Hymenium: Basidia 30-46 × 10-15 µm, hyaline to yellowish, some slightly pseudoamyloid, all 4-spored or (1-2)4-spored. Cystidia

(17)35-75(97) × 6-10(20), versiform, often mucronate, most frequently ampullaceous to fusoid or ventricose, but not distinctly dimorphic, hyaline or more often with yellow, melleous or fuscous contents (pseudocystidia) but inamyloid (the content remaining unchanged in Melzer's reagent), the apex if present 9-11(42) μm long, thin, obtuse, sometimes separated by a secondary septum. Cheilocystidia sometimes moderately thick-walled and much more often clavate than pleurocystidia.

Hyphae without clamp connections, inamyloid. Hymenophoral trama bilateral. Epicutis of pileus (between the white rimulosities if present) a trichodermial palisade, with a tendency to produce an interwoven layer, with few somewhat inflated or short cells, these 8-14 μm broad, with obtuse upper tip, some internally granular in Melzer, mostly hyaline in KOH. Endocystidia often present in hymenophoral trama but scarce.

Chemical color reactions: NH_3 slowly becoming purple violet on the pileus surface.

Under conifers (*Pinus* and *Abies*) on the ground or litter.

Material studied: MEXICO, Michoacán: km 45 of the road Morelia-Cd Hidalgo. 21 VII 1983. García 3697 (F, ITCV). — Veracruz, E of Cofre de Perote, El Revolcador, 4 VIII 1983. Villareal 5.8 (F, XAL).

This *Porphyrellus* is known from the Pacific northwest of USA and is new for Mexico. It keys out with *T. pacificus* in Wolfe's key and fits his description reasonably well. We have not studied the type from Washington. We describe only the Mexican specimens.

2. **Porphyrellus umbrosus** (Atk.) comb. nov. (Pl. 17, fig. 51; Pl. 8, fig. 58-61).

Syn.: *Boletus umbrosus* Atk. Journ. Mycol. 8: 112, 1902.
Tylopilus umbrosus (Atk.) Smith & Thiers. Mycologia 60: 950. 1968.

Pileus leather brown, dry, finely rimulose-areolate, with cream color context showing between areolae, convex, about 37 mm broad.

Hymenophore tubular; tubes cream colored, adnate or adnexed, pores whitish cream color, reddish spotted, round to angular, small (mostly smaller than 1 mm wide), eventually brown where bruised like the tubes. Spore print not obtained, vinaceous brown according to Wolfe (1979).

Stipe pale gray, older or on drying becoming fuscous ("Madrid" to "Olive wood" M&P), subglabrous, not reticulate, slightly longitudinally striate with pale gray, subequal, tapering downwards into a short pseudorrhiza, 35 × 11 mm; basal mycelium whitish to greyish white; velum none.

Context white, turning blue in tube walls and above the hymenophore on injury, odor and taste mild.

Spores (10)10.8-13.3(15) × (4.8)5.2-6.3 μm ellipsoid to ovoid or subfusoid, mostly with suprahilar applanation or a slight depression, more frequently narrowed above than broadly rounded, in some reddish orange in Melzer, in others subhyaline to brownish as in KOH, with refractive olive-pallid guttulae inside, those not visible in the Melzer. Outer wall dark brown in most mature spores, not distinctly pseudoamyloid, but young spores cyanophilous.

Hymenium: Basidia 32-52 × 9.5-12.5 μm, often pedicellate, 4-spored. Cystidia 40-54 × 8-18.5 μm fusoid, mostly hyaline and some granular inside, not pseudoamyloid (pale yellow in Melzer). Cheilocystidia 30-50 × 12.5-17 μm, clavate to clavate-mucronate, pedicellate, not incrusted, otherwise like pleurocystidia.

Hyphae without clamp connections, inamyloid to vaguely pseudoamyloid in part. Hymenophoral trama bilateral of the *Boletus*-type, with hyaline hyphae 3.5-11.5 μm diam. Some oleiferous hyphae 4.5-10 μm broad.

Epicutis of pileus in the areoles a trichodermial palisade which becomes irregular, with hyphae mostly cylindrical to narrowly ventricose, with a few subisodiametrical ones intercalated, the terminal cells often cystidioid, 29-50 × 6-10 μm, cylindrical or widened above or below, with obtuse apex, thin-walled, often with very finely granular contents, only in superposition very slightly orange-ferruginous, definitely not gelatinized.

Chemical characters: Not checked.

In *Quercus* woods on the ground or litter, on hills.

Material studied: MEXICO: Chiapas, San Cristobal de las Casas, Cerro del Huitepec, 18 VIII 1987. García 5461 (F, ITCV).

This species fits the description of American material and the type by Wolfe (1979) and is easily identifiable in Mexico where it is geographically and ecologically separable from the other species with inamyloid cystidia. It is also smaller than the other Mexican species and has a pseudorrhiza which is inserted in the litter but this pseudorrhiza is not perennial nor acute below but rounded. Atkinson in his original description has not noticed the bluing nor the blue ring Wolfe (1979) mentions. This color ring at the apex of the stipe is very fugacious and was not noticed by us either. It is more noticeable but not quite constant in *P. cyaneotinctus* which differs in pseudoamyloid hymenial cystidia and larger size. *T. sordidus* and *T. snellii* differ in longer spores (see notes to the following species).

3. **Porphyrellus cyaneotinctus** (Smith & Thiers) Sing. comb. nov. (Pl. 5, fig. 62; Pl. 19, fig. 63-65; Pl. 10, fig. 66-69).

Syn.: *Tylopilus cyaneotinctus* Smith & Thiers, Mycologia 60: 952. 1968. *Porphyrellus pseudoscaber* ssp. *cyaneocinctus* Sing. sensu Singer, Farlowia 2: 116. 1945 (synonymy excl.).

Pileus ochraceous brown to mostly (''Saccardo's umber'' to ''snuff brown'' Ridgway), often with leather brown or fuscous portion, velutinous to subvelutinous, cracking readily into areoles so that all adult specimens are conspicuously rimulose-tesselate, with the crevasses whitish, rarely bluish, running in all directions, not viscid, convex, 25-115 mm broad.

Hymenophore tubular, yellowish white to grayish white (between ''cream buff'' and ''cartrige buff'' Ridgway), ± deeply depressed around the stipe and eventually with a tinge of chestnut brown from accumulated spores, greenish blue on injury, leaving no sterile margin outwards; tubes up to 21 mm long, convex on lower (pore-) surface in mature specimens; pores concolorous with the tubes, and discoloring like these, then becoming brownish red or at least dirty pinkish, angular at least in mature specimens, 0.6-1 mm wide, but rather wide in older specimens raching 2 mm diam. Spore print ''warm sepia'' in a good print.

Stipe whitish-yellowish at the very apex with lamelloid proliferations running down from the wall of the tubes for a few millimeters, to a characteristic though sometimes indistinct green ("deep lichen green" to "rejane green" Ridgway) color ring, rarely without this, but the ring on drying and when older disappearing, below the color ring concolorous with the pileus later becoming "Saccardo's umber", subequal, usually tapering upwards, solid, 46-70 × 7-16 mm; basal mycelium white or whitish; veil none.

Context white in the pileus, under the cuticle often turning violet, otherwise all or in portions blue on injury ("porcelain blue" or "gobelin blue" Ridgway) near tubes but almost entirely bluing when young and fresh, the expressed sap coloring white paper blue, also ± bluing on the apex of the the stipe; odor none or very slightly of overmature fruit; taste mild.

Spores 10.5-14.3 × (4)5.5-7.2 μm, in some prints occasional but rare giant spores to 18 × 9 μm, versiform in shape but usually fusoid, also ellipsoid-oblong, with or without a suprahilar depression or applanation, more rarely without it, usually with a wall consisting of a brownish inner layer and hyaline outer layer, which looks like an optical halo, smooth, exceptionally vaguely punctate, occasionally rather thick, especially in the apical, rarely the basal portion, (to 0.8 μm), as a whole yellowish to cinnamon and often with an olivaceous tinge, in KOH, many weakly pseudoamyloid (pale and dirty ferruginous brown to reddish brown in the Melzer), without a germ pore.

Hymenium: Basidia 24-51 × 9.5-14.5 μm, 4-spored. Cystidia 45-72 × 7-20 μm, fusoid to ventricose-mucronate, but with ± obtuse tip, hyaline to brownish or brown in KOH, with guttulate contents, rarely without it, the contents mostly dark (yellow-) brown in the Melzer. Cheilocystidia 48-72 × 14-22.5 μm, mostly clavate and pedicellate, thin-walled, some becoming somewhat thick walled later, with some transitions in form to the (pleuro-)cystidia in some specimens.

Hyphae without clamp connections, some oleiferous hyphae present (3.5-10 μm diam.), some becoming orangy or subferruginous in the Melzer. Hymenophoral trama bilateral of the *Boletus*-type, the mediostratum at first darker and denser than the lateral stratum, soon all hyphae equally pigmented and aspect more similar to the *Phylloporus*-type, hyphal diameter 4.2-12 μm; subhymenium narrow.

Epicutis a trichodermial palisade at least in the outermost layer when young, below and in age with somewhat interwoven hyphal chains, the terminal cells dermatocystidial rising from the lower layer of sub-interwoven hyphal chains, 26-51 × 7-21 μm, fusoid, the apex ± rounded, hyaline in KOH, deep ochraceous to ochraceous orange in the Melzer, the lower cells mostly ± elongated but usually some cells very short, even the terminal cell sometimes short, the cells 8-16 μm broad. Covering of the stipe with numerous dermatocystidia, these mostly like the cheilocystidia.

Chemical characters: NH_3 and NH_4OH on pileus deep purple (''dark livid brown'' Ridgway), eventually ''Engish red'' (Ridgway); on the context yellowing, then often greening, in some part of the context merely indefinitely sordid or with little reaction; on the pores often bluing; on the stipe orangy with violet periphery; — KOH on the pileus as with NH_3 but less bright, often more red-brown' on context often greenish, but after bluing ''Etruscan red'' becoming ''vinaceous tawny''. — $FeSO_4$ on context ''deep glaucous gray'' or ''dark glaucous gray'' (Ridgway). — Formol on context salmon. — Methylparamidophenol: negative.

On the ground and litter under *Quercus*, for example forming ectomycorrhiza with *Quercus rysophylla,* in Florida with *Q. virginiana* and *Q. laurifolia,* in Mexico mostly in mesophytic woods.

Material studied: MEXICO: *Tamaulipas:* Rancho ''El Cielo'', 4 X 1985, 1150 m alt. García 4875 (F, ITCV). — *Nuevo León:* El Ranchito 600 m alt. 21 VIII 1980, García 156 (F, ITCV). — Same place, 26 V 1980, García 154 (F, ITCV). — El Alamo, alt. 550 m, 7 VII 1984, García 4466 (F, ITCV). — Cañon de Puerto Genovevo, 1000 m alt., 4 X 1982, García 2888 (F, ITCV). — Joya del Venado, Sierra de Chipinque, 24 IX 1983, alt. 850 m, García 3166 (F, ITCV). — Also various collections from Florida (see Singer 1945).

The first, exact and complete description of this species was published as *Porphyrellus pseudoscaber* spp. *cyaneocinctus* Sing. (1945, p. 20). However Singer considered the species of this group as subspecies at this time and identified it with the type of *Boletus fumosipes* Peck. In the first place, these subspecies are now considered to be species although very closely related to each other and the type of *B. fumosipes* has been designated by Wolfe & Petersen in 1978; according to

Wolfe's (1979) type studies, this is not identical with my specimens from which the description of ssp. *cyaneocinctus* was made up. Looking for a new name we used the key to (sg.) *Porphyrellus* published by Wolfe and were lead to *Boletus fumosipes* again.

In the meantime, we found that Smith & Thiers had published our subspecies as *Tylopilus cyaneotinctus* (!) exhibiting the same characters we emphasize including the nondimorphic, at least partly pseudoamyloid hymenial cystidia, the relatively small spores and the broad epicuticular hyphae, the blue color ring at the apex of the stipe. This was confirmed by Wolfe's (1979) study of the holotype. As much as we can judge by the description, we are inclined to think that *T. fumosipes* var. *glabripes* Wolfe is again the same species.

Smith & Thiers (1968) did not see (or were unsure about) the similarity of *Porphyrellus pseudoscaber* spp. *cyaneocinctus* and *Tylopilus cyaneotinctus,* perhaps because they mistakenly stated that my specimens came from Austria (p. 954) whereas all my Austrian specimens had smooth spores (Singer 1966) and had nothing to do with the Floridian specimens described in 1945. And my notes on *P. sordidus* came from the lectotype (FH), also unknown in Austria. Or did they erroneously identify my Florida species, the one her described, with *Boletus sordidus,* just the same as they misinterpreted *B. fumosipes* as *B. sordidus*?

Boletus sordidus Frost is a species with dimorphic cystidia and has therefore been referred to *T. porphyrosporus* as var. *sordidus* by Wolfe; ist has, however, different spores according to the type studied of both Singer (1945, p. 119) and Wolfe (1979, p. 121) not only from *P. porphyrosporus* var. *prophyrosporus* but also from those of *P. cyaneocinctus.* In addition, it is considerably smaller than *P. porphyrosporus.* In Wolfe's key (l.c. p. 78-79) *B. sordidus* is indeed keyed out twice, once under under the species with dimorphic pleurocystidia, and once with considerably shorter spores than in his description of var. *sordidus* (p. 54). In my own type studies on *B. sordidus* are found the broadest spores with walls 0.5 and up to 1.7 μm thick, some with incomplete germ pore or subtruncate. A few exceptional spores were up to 18 μm long; in one collection very exceptional spores were found with spore ornamentation of type XI (but it is barely possible that these were foreign). Halling (1983) found that specimens from Vermont, USA, agreed with the analysis by Wolfe (hence pleurocystidia

has, only for this reason been included inasmuch as the species has been reported from Mexico by Welden & Guzmán (1978) from the same state of Veracruz, mostly from localities of high altitudes (first mention of this species in Mexico) and was specifically mentioned as occurring on the Cofre de Perote Ver. by Guzmán & Villareal (1984). If this species, theirs and ours, corresponds rather to *B. fumosipes* Peck according to the available microscopic characters, we would be inclined to subordinate it to *P. porphyrosporus* as some intraspecific taxon. But *B. sordidus* (see also p. 67) should likewise be compared.

V. FISTULINELLA Heim

Englers Bot. Jahrb. 30: 43. 1901.

Syn.: *Ixechinus* Heim, Rev. Mycol. 4: 20. 1939 ex Heim l.c. 30: 233. 1966.
Mucilopilus Wolfe, Mycotaxon: 117. 1979.

The genus is here understood in a wider sense, including one species with pseudocystidia, one with initially dry pileus associated with Pinaceae, as well as Wolfe's genus *Mucilopilus,* here considered as a section of *Fistulinella*, but all sections characterized by an ixotrichodermium which mostly exists in all stages, or, in one section, develops with the maturing of carpophores. On the other hand, we temporarily exclude *Boletus mucosus* Corner (1972, p. 77) because of its rough spores, although its viscose surfaces, refer to *Fistulinella*. The sections keyed out below are partly genera according to Wolfe, and may be elevated to subgeneric status once all the paleotropical taxa are fully known.

Key to the sections of Fistulinella represented in the region

1. Pileus at first dry, developing an ixotrichodermium while maturing............ .... Sect. *Scrobiculatae*

1. Ixotrichodermium present even in young specimens; not associated with Pinaceae.

   2. Spores 9.5-11(12) × 4-5.2 μm or × 3-3.7 μm. On the ground.............. .... *F. jamaicensis; F. guzmaniana*

   2. Spores larger, at least some over 11 μm long.

      3. Pseudocystidia numerous in hymenium and on the surface of stipe....... .... Sect. *Wolfeanae*

      3. Pseudocystidia none............................. Sect. *Fistulinella*

Section SCROBICULATAE (Sing.) Sing. Agar. Mod. Tax. 4th ed. p. 798. 1986.

Syn.: *Scrobiculati* Sing. (as. sect. of *Tyloplilus*), Am. Midl. Nat. 37: 95. 1947.
Scrobiculati (Sing.) Sing. (sect. of *Porphyrellus*), Agar. Mod. Tax. 3d ed. p. 748. 1975.
Scrobiculati (Sing.) Wolfe, (as sect. of *Mucilopilus*) Mycotaxon 108. 1979.

Pileus initially dry, often scrobiculate. Ectomycorrhiza with *Pinus* or *Quercus*. Type species *B. conicus* Rav. ap. Berk. & Curt.

Key to the species

1. Pileus fresh scrobiculate, yellow to rusty brown; pores ± vinaceous to pinkish, less than 1 mm wide. Ectomycorrhiza with *Pinus*.
   2. Pileus pulvinate to convex, obtuse; stipe glabrous or almost so; context negative to partly weakly rosascent with NO_4OH.......................... .... 1a. F. CONICA var. CONICA
   2. Pileus with a distinct umbo; context in very fresh condition reddening with NH_4OH; stipe thinly tomentose above and finely pustulate-scurfy below.. .... 1b. F. CONICA var. BELIZENSIS
1. Pileus more fuscous on gilvous ground, not distinctly scrobiculate when fresh; pores when fresh olive to grayish olive, up to 1 mm wide. Mycorrhiza with frondose trees, apparently *Quercus,* and *Monotropaceae*....... 2. F. ALFAROAE

1a. **Fistulinella conica** (Rav. ap. Berk. & Curt.) Pegler & Young, Trans. Brit. Myco. Soc. 76: 140. 1981.

Syn.: *Boletus conicus* Rav. ap. Berk. & Curt. Ann. Mag. Nat. Hist. 12: 430. 1853.
Suillus conicus (Rav. ap. Berk. & Curt.) Kuntze Rev. Gen. Pl. 3(2): 535. 1898.
Tylopilus conicus (Rav. ap. Berk. & Curt.) Beardslee, Mycologia 26: 253. 1934.
Porphyrellus conicus (Rav. ap. Berk. & Curt.) Sing., Agar. mod. tax. 3rd ed. p. 748. 1975.
Mucilopilus conicus (Rav. ap. Berk. & Curt.) Wolfe, var. *conica.*

Pileus with uneven surface of elevated ridges around small depressions (scrobiculate) that give the pileus a characteristic pitted appearance, the most elevated ridges "Sudan brown," "raw sienna," "yellow ocher," (Ridgway) the low ones "mustard yellow," the depressions (since they are naked context) white, but the pileus giving the general impression of about "primuline yellow", more pitted near the margin, pulvinate to convex, 25-95 mm broad.

Hymenophore "pale grayish vinaceous" to "pale vinaceous fawn" when mature, paler when immature, depressed around the stipe; tubes about 14 mm long; pores small, equal, concolorous and unchanging, 8-10 per 5 mm radially, 10-11 per 5 mm transversely; spore print "Rood's brown."

Stipe white, at least at the apex and at the base, the middle portion usually with a flush of "light pinkish cinnamon" and "chamois," smooth to minutely lineate vertically or rugulose at places, entirely glabrous, ventricose to subequal, 40-80 × 6-20 mm; mycelium white.

Context pure white with a hyaline line above the tubes; odor agreeable, fruity; taste mild.

Spores 14.2-18.2(21) × 4-6 μm, fusoid, the upper half attenuate, with thin, rarely in some exceptional spores with slightly thickened walls (0.8 μm), hyaline and becoming melleous or brownish melleous only at maturity, with distinct or indistinct suprahilar depression, with small oil-drops inside, smooth.

Hymenium: Basidia (17.5)23.5-36 × 10-11.6 ×, 4-spored; cystidia 31-68 × 4.8-9.5 μm subfusoid, often subacute, often with 1-2-(3) septa, hyaline but frequently with a melleous incrustation.

Hyphae: Hymenophoral trama truly bilateral of the *Boletus*-type. Epicutis formed by strands of very slender, and others of medium thick (4.5-8 μm in diameter) hyphae; these hyphae are alternately hyaline and deep honey color with dissolved pigment, somewhat interwoven and repent but a few ascendant to erect, all without clamp connections.

Chemical reactions. — KOH on surface of pileus almost negative to somewhat deeper colored, eventually brown; on context of pileus ± yellow, eventually brown; on context of pileus, negative except for the margin which usually assumes the color of the tubes. H_2SO_4 on pores, negative. — $FeSO_4$ on context of pileus, negative but if treated with KOH subsequently, the context becomes brownish; on tubes with $FeSO_4$, grayer, eventually steel gray. — Formol on surface of pileus and on context, negative. — Methylparamidophenol on context and tubes, negative. — Phenol and NH_4OH on most parts almost or quite negative.

In flatwoods and moist mixed woods under *Pinus,* e.g. *Pinus palustris* sometimes on soil that is at times partly inundated. Fruiting in July to September.

Distribution. — From South Carolina to Florida, Texas, and Mexico, but not in the tropical zone.

Material studied: — MEXICO, Guerrero, Mun. Mochitlán, Agua del Obispo, 1 VII 1982, alt. 950 m. L. Hernandez s.n. (F, FCME). — Capello 393 (F, FCME). — Also material from USA: North Carolina, Florida, and Texas

Var. **belizensis** Sing., Araujo & Ivory, Beih. Nov. Hedw. 77: 140: 1983. (Pl. 23, fig. 75A, 75B).

Pileus straw color to pale luteous, with yellowish fulvous fibrillose scales, these appressed and unshining on shiny ground, scrobiculate, conic-convex and umbonate, up to 55 mm broad. — Hymenophore tubular, the tubes long (to 8 mm), pinkish, with the pore surface ventricose-convex, adnate to slightly depressed around the stipe; pores concolorous or rosy-buff, angular but subisodiametric and small (about 0.5 mm wide when dried), only at the stipe narrowly sublamellar-elongated. Spore print between vinaceous buff and fawn when fresh, dehydrated still about "terrapin" (Maerz & Paul). — Stipe stramineous with a pale luteous to yellow zone near base, the base itself white or whitish, at the apex thinly tomentose, scurfy-pustulate below (slightly so, distinctly so under a lens), dry solid, subequal, up to 60 × 10 mm. Basal mycelium white. Veil none. — Context white, unchanging, firm-fleshy. Odor slight. Taste mild or sweet.

Spores (10.5)13-17 × 3.7-4.8 µm, most frequently (from print) 13.5-14.5 × 4-4.4 µm, with suprahilar depression, fusoid, fewer oblong to cylindric, smooth, without germ pore, pale smoky-melleous in superposition and then slightly pseudoamyloid, singly in KOH yellowish to subhyaline and seemingly inamyloid, cyanophilic.

Hymenium: Basidia 18-27 × 7-9.5 µm, mostly 4-spored. Csystidia on edge and side similar, moderately numerous, 22-40 × 6.5-10.7 µm, fusoid to ampullaceous, hyaline, thin-walled, without pseudoamyloid granulation or any other enclosures.

Hyphae inamyloid, without clamp connections. Hymenophoral trama bilateral of the *Boletus*-type, with subhyaline mediostratum consisting of 3-4 μm broad hyphae, the lateral stratum gelatinized of strongly curved (outwards) hyaline hyphae 2-3 μm broad, in older tissue less divergent and hyphae reaching 6 μm diam.

Cortical layers: Epicutis of pileus - an ixotrichodermium of subhyaline elements which, in the outermost layer become somewhat depressed with sometimes branched or nodose and more strongly gelatinized, rather narrow terminal cells, but these elongated, below this (hypodermium) passing into a pigmented layer of partly broader hyphae which are less gelatinized and brown from an intracellular pigment which becomes slightly more chestnut in Melzer's reagent, but without incrustations. Covering of the stipe consisting of a monostratous layer covering in places verruculose elevations of the stipe trama continuous at the apex and discontinuous (as in *Conocybe*) further downwards; this monostratous covering consisting of (1) dermatobasidia, these about 17 $\times$ 6.5 μm and often 2-spored, (2) bodies resembling sterile basidioles, these 13-19 $\times$ 4-8 μm, hyaline, ventricose to clavate or subclavate, very numerous, (3) dermatocystidia like the hymenial cystidia, e.gr. 25 $\times$ 7 μm with the apex 2 μm across, hyaline, fusoid to ampullaceous, thin-walled, less numerous.

Chemical color reactions: Context rapidly reddening with NH_4OH.

In small groups under *Pinus caribaea* with ground cover of grass and few herbs and shrubs.

Material studied: BELIZE: Augustine Forest Station 17 VI 1976. Ivory S/114 (F).

F. conica var. *conica* and var. *reticulata*, the latter not (yet?) known from this region, differ by not umbonate pileus and often reticulate or sub-velutinous stipe, by the context not reddening with NH_4OH, and slightly broader spores.

2. **Fistulinella alfaroae** Sing. & Gómez spec. nov. (Pl. 22, fig. 76; Pl. 24).

Pileus light fuscous or fuscous on gilvous ground, under a lens fuscous brown web-like fibrillose on glabrous yellowish ground, dry, subviscid in age when wet smooth, pulvinate to convex, 34 mm broad.

Hymenophore tubular. Tubes long, olive to greyish olive, deeply depressed around the stipe and there lamellar-extended; pores concolorous, many (not all) compound, with uneven pore surface, but not showing separable pores as in *Fistulinella*, up to 1 mm wide, unchanging when touched or bruised. Spore print purple pink when fresh, dried brown (15 J 11, M&P).

Stipe reddish brown, or fulvous to almost fox red, or cinnamon brownish, smooth and glabrous, naked, dry, solid, often curved, equal to very slightly and gradually tapering upwards, 82 × 4-4.5 μm; veil none; basal mycelium pure white, with white rhizomorph-like formations; pseudorrhiza none.

Context white or whitish in pileus, unchanging or very slightly sordescent but in worm-holes becoming slowly reddish; in the stipe concolorous with the surface, dirtier than surface when dried. Taste slightly acidulous, not bitter. Odor none.

Spores 12-16(18.5) × 4-5.5(6.2) μm, mostly 12.5-15.5 × 5-5.5 μm, giant spores 16.5-18.5 × 5-5.5 μm, not truncate, light melleous in KOH, with moderately to slightly thickish wall (e.g. 0.5 μm thick, episporium dark and overall pale livid in Melzer, (i.e. subinamyloid), in light microscopy quite smooth, with SEM subsmooth with few shallow scrobiculations, without germ pore.

Hymenium: Basidia 19.7-22 × 8.5-13 μm, 4-spored. Cystidia on pores and inside tube 35-48 × 5-7 μm ampullaceous, apex 3-4 μm broad, sometimes twice narrowed, the tip obtuse, quite hyaline and without any optically visible contents, not pseudoamyloid, cheilocystidia and pseudocystidia not differentiated.

Hyphae inamyloid, without clamp connections. Hymenophoral trama bilateral of the *Boletus*-subtype, but somewhat transient to the *Phyllosporus*-subtype since the mediostratum is hardly more pigmented than the the totally hyaline lateral stratum which is but moderately gelatinized; hyphae here of variable diameter.

Cortical layers: Epicutis of pileus (the fuscous fibrils) of appressed strands of parallel hyphae 3-10 μm in diameter, filamentous or cylindrical, not gelatinous but the walls slightly thickened and superficially gelatinizing and becoming very thin and sometimes undulating in the process, hyaline or with ochraceous intraparietal pigment, some thinly granularly incrusted. Hypodermium a trichodermium of hyphal elements which are scarcely gelatinized where it is exposed to the surface, but non-gelatinized otherwise, pale yellowish, or yellowish-hyaline, very different in length and diameter, running in strands in different directions. Covering layer of stipe only slightly and intermittently gelatinized, hyphae in the outermost tier thin-filamentous and often undulating, thin-walled and hyaline in KOH everywhere, dermatocystidia and dermatopseudocystidia not present.

Chemical characters: NH_4OH on pileus surface fulvous with a blue zone around the drop. — NH_3 little reaction — $FeSO_4$ slowly greenish gray on context — KOH on surfaces negative.

In mixed montane forest of the lower altitudes, growing in association with a population of *Monotropa* in small groups under oaks.

Material studied: COSTA RICA: Cartago, Valle de la Estrella, between 1400 and 1500 m altitude, 5 VIII 1986. Leg. Singer, Gomez and Alfaro, Singer B 14621 (F), type.

This species, obviusly not identical with any described one, has the spores of a *Porphyrellus,* and the cystidia of a *Fistulinella*. But in spite of the almost dry pileus we believe it belongs in *Fistulinella*. We insert it in *Fistulinella* because its habit, its tendency to gelatinize, the absence of pseudocystidia and the large spores exclude *Tylopilus.* There is no precedent for the assumption of a smooth-spored group in *Austroboletus* or a pigmented one without veil in *Veloporphyrellus.*

Section PSEUDOTYLOPILI (McNabb) Sing. comb. nov.

(Basionym: *Porphyrellus* sect. *Pseudotylopili* McNabb, N. Zeal. J. Bot. 5: 546. 1967).

Type species: *Porphyrellus viscidus* McNabb (*Fistulinella viscida* (McNabb) Sing. Persoonia 9: 435. 1978).

This section which contains the type species of the original concept of *Mucilopilus* Wolfe is at present not known from our region. However, a species which might belong in this section, *F. jamaicensis* (Murr.) Sing. has been described from Jamaica (W.I.) in young growth of coppice among grass. Similar material has been collected in Texas

(U.S.A.). The species is not fully described, but may eventually be observed in Mexico. A description may be found in Singer. Araujo & Ivory (1983, p. 142), but the presence of pseudo- and cheilocystidia was not established nor was its habitat ecologically analyzed. *P. brunneus* McNabb may also belong here.

Section FISTULINELLA

This is the section containing the type species of *Fistulinella* Henn viz. *F. staudtii* Henn. It was analyzed by Singer (1973, p. 315). It shows complete coincidence with the characters of *Ixechinus* Heim. To emphasize the surface of the carpophores as described in specimens preserved in alcohol for 72 years where the gelatinized layers have deteriorated does not seem realistic inasmuch as the pore edges have evidently likewise deteriorated to the point of showing free hyphal ends. The free separable tubes are characteristic of this section, as is the lignicolous habitat and thin stipe. Even if a slight punctation which is now visible under a good lens should indicate some truly trichodermial portions on the surface, it would not be in contradiction with the basic characters of the section.

Pileus viscid and tubes free at pore level with naked, macroscopically glabrous, smooth stipe and white pileus; cheilocystidia absent or not recorded, either not forming ectomycorrhiza or not forming it with Pinaceae or Fagaceae, but possibly associated with Sapotaceae(?). Polygonaceae(?), Euphorbiaceae(?), and other Dicotyledones, mostly on rotting wood, tropical Africa and America.

3. **Fistulinella mexicana** Guzmán, Bol. Mex. Mic. 8: 54. 1974 (Pl. 22, fig. 19).

Syn.: *Muciloporus mexicanus* (Guzmán) Wolfe, Mycologia 74: 39. 1982.

Pileus yellow-café, or whitish, viscid, glabrous, irregularly areolate, convex, then applanate, 35 mm broad.

Hymenophore tubular, tubes free (in formalin), pale brownish-yellow, somewhat adherent to the stipe, separable from the context, 6-8 mm long, pores concolorous, circular or sometimes subcircular, rarely anastomosing, 1-1.5 mm wide.

Stipe white, viscid, slightly fibrillose, cylindric, with slightly attenuated base which is disciform, covered at the base by a trasnparent vagina which is very glutinous, 75 × 6 mm. Context fleshy-gelatinous.

Spores 11-14.2 × 4.3-5.3 μm, pale melleous in KOH, oblong to fusiform, smooth, with suprahilar depression, most with indistinct to distinct germ pore because of an interruption of the episporium at the distal end, inamyloid to weakly pseudoamyloid (seen to be so only in superposition), some vaguely tilda-shaped or subreniform.

Hymenium: Basidia 25-29 × 8-10.5 μm, 4-spored. Cystidia present but scanty in amyloid, according to Guzmán "24-30 × 5-6 μm, cylindric withthe apex slightly swollen". Cheilocystidia none (Wolfe). Hyphae hyaline or with slightly melleous interior, inamyloid, without clamp connections. Hymenophoral trama longitudinally ruptured, entirely gelatinous with the outermost hyphae tending to ascend, obviously rest of a bilateral structure.

Cortical layers: Cystidiform hyphae, 35-48 × 12-16 μm cylindric-globose, hyaline (Guzmán) on the surface of pileus and stipe; ixotrichodermial on pileus (Wolfe).

Solitary, lignicolous in "selva alta subperennifolia".

Material studied: MEXICO: Campeche, 10 km S of Escárcega, 17 XI 1971, Salgado-Baena (ENCB), typus.

The above description is adapted from the description by Guzmán who studied material preserved in formalin part of which we could study. Most of the microscopical description (except where his observation is quoted) is ours. Whether there is any ectomycorrhiza associated with this remains unknown. The forest contains species of many genera of trees which are enumerated by Guzmán. His conclusion that his species proves floristic relations between Mexico and Africa, although such relations certainly exist, is somewhat flawed by the fact that related species have been observed also in South America, North America, New Zealand, and apparently also in tropical Asia. The glutinous "vagina" described by Guzmán makes one think of *Boletus mucosus* Corner (1972).

F. mexicana differs from the Madagascarian *F. minor* and *F. major* (Heim) Guzmán in being larger respectively smaller, having somewhat different spore measurements, and growing on wood rather than humus. Also, the colors and the size and shape of the cystidia seem to be slightly different in the African species. *F. staudtii* Henn. differs in much larger spores (15-20 × 4.5-6.2 μm), and narrower pores (about 1/4 mm). The carpophores are lignicolous as in *F. mexicana.*

F. gloeocarpa Pegler (Pegler 1983) differs only slightly from *F. mexicana* in somewhat narrower pores and somewhat longer spores and might be the species most closely related to *F. mexicana.*

Section WOLFEANAE sect. nov.

Pseudocystidiis praesentibus. Typus: *F. wolfeana* Sing. & García.

4. **Fistulinella wolfeana** Sing. & García nov. spec. (Pl. 22, fig. 78).

Pileus pink, viscid, minutely shallowly scrobiculate, convex, 70 mm. Hymenophore tubulose, tubes whitish-cream, unchanging, at stipe depressed; pores concolorous, 0.5 mm wide, when injured ± ochraceous. Spore print not obtained.

Stipe light yellowish brown, finely reticulated, but reticulum consisting of rows of minute brown punctations, solid subbulbous, 50 × 18 (at base) mm.

Context yellowish white, fleshy, unchanging when injured; odor mild, fungoid; taste mild.

Spores 9-13.5(14.3-17) × 4.5-5.5 μm, most frequently 9.5-13 × 4.7-5 μm, subfusoid, often broadest in lower third, without suprahilar depression but sometimes with suprahilar applanation, smooth, melleous hyaline, interior mostly weakly pseudoamyloid (pale dirty livid in Melzer), wall hyaline or subhyaline

Basidia ± 22 × 8.5 μm, 4-spored. Cystidia and pseudocystidia present, 24-40 × 9-12 μm, but also larger, in tubes most filled with amorphous chestnust bodies or granulations in Melzer, others not, the contents also visible in KOH but there yellowish or subhyaline, some material intercellular-subincrusting; cheilocystidia numerous, similar to (pleuro-)cystidia, also pseudocystidioid as those, but in an average smaller e. gr. 20 × 6.5 μm.

Hyphae without clamp connections. Hymenophoral trama yellowish melleous, mediostratum consisting of scarcely gelatinized, interwoven hyphae, lateral stratum rather strongly gelatinized, hyaline and diverging (bilateral of the *Boletus* type).

Epicutis of the pileus an ixotrichodermium which is rather deep, with hyphae 2-4 μm diam., hyaline, some melleous incrusted, some hyphae dermato-pseudocystidioid as the hymenial pseudocystidia. Hypodermium differentiated by being more pigmented, more granulated and partly non-gelatinized or less gelatinized, the hyphae 2.8-6 μm broad. Both layers in Melzer yellow-brown to tan. Covering of the stipe in the reticulated part consisting of dermatocystidia and dermatopseudocystidia, the former e. gr. 23 × 5.5 μm, ventricose or ampullaceous or mucronate, the latter even more polymorphous, with inside bodies like the hymenial cystidia and 26-33 × 4.5-11 μm.

On the ground in a mixed wood of *Pinus* and *Quercus*, associated rather with *Quercus,* but ectotrophic mycorrhiza not demonstrated.

Material studied: MEXICO: Hidalgo, Minas Viejas, at 2000 m, 31 VII 1981. García 658 (F), type. — México: Mun. Tejupilco, Nanchititla, 200 m alt. Kong Luz 423 (F, ENCB). — Gonzales-Velasquez, 26 VIII 1988, 806, 718 (F, ENCB).

This species has all the characters of *Fistulinella,* but differs in the presence of pseudocystidia and dermato-pseudocystidia. It seems that *F. nipponica* Nagasawa in Sing. belongs to this section although there the pseudocystidia are much weaker reacting with the Melzer, as shown in topotypical material collected by Nagasawa & Singer, No. A4009 (F).

This section and species are named for Dr. C.B. Wolfe, Jr. whose studies in the *Tylopilus-Fistulinella* (*Mucinopilus-Ixechinus*) group were quite extensive and useful.

Tylopilus subfusipes A.H. Smith (Persoonia 7: 330, 1973) seems close according to the description but has narrower spores, hymenophore not unchanging, and bitter taste.

5. **Fistulinella guzmaniana** nob. ad int. (Pl. 22, Fig. 77).

Pileus reddish, smooth, "dry" subglabrous, convex, obtuse, 16 mm broad when dried.

Hymenophore tubulose "whitish", pores concolorous, up to 0.9 mm wide when dried, some isodiametric, some elongated, appearing sub-lamellate near stipe. The tubes appear slightly non-geopetalous, the pores vary widely in diameter, some are compound. Spore print not obtained, the accumulated spores on the hymenophore when seen under a lens in dried specimens appear ocher-pallid.

Stipe "red-brown" in the middle, white at apex, finely tan flocculose on paler ground, dried yellowish pallid with a dull rosy-red shade in the middle, solid, cylindrical, dry, 48 × 4 mm when dried. Basal mycelium white. Context white, unchanging, but there are traces of a slight browning where the pores were touched. Mild. Odor not described.

Spores 8.2-10.5 × 3-3.7 µm, fusoid to cylindric, smooth, with suprahilar depression, without germ pore, hyaline to subhyaline in KOH.

Hymenium: Basidia 23-29 × 7.5 µm, 4-spored. Cystidia fusoid, with protracted obtusate or acute tip, with firm, brown or hyaline wall 0.2-0.4 µm thick, with or without amorphous contents bodies, numerous.

Hyphae without clamp connections, in the pileus mostly filamentous but some widened (up to 11 µm), strongly interwoven, subgelatinous, hyaline. Hymenophoral trama bilateral of the *Boletus*-type, mediostriate brownish, intraparietal pigment, axially arranged and not gelatinized; lateral stratum broad an with strongly recurved hyphae 5-6.5 µm broad and gelatinized.

Cortical layers: Epicutis of pileus of interwoven hyphae 2-7 µm broad, its upper layer tending to become gelatinized, below not gelatinized and hyphae firm-walled but otherwise similar, not incrusted (or scarcely so), hyaline to yellowish and in places slightly brownish (dried material, in KOH).

In "bosque caducifolio con *Liquidambar*".

Material studied: MEXICO: Veracruz: SW of Banderilla, Cerro de la Martinica, 1500 m alt., 10 VII 1976 (INPA).

The position of this species is still doubtful and its insertion in sect. *Wolfeanae* is tentative. The cuticle of the pileus is much like that of sect. *Scrobiculatae*, but the spores are small, and remind one of *F. jamaicensis* (with broader spores and without inclosures in the cystidia). The macroscopical structure of the hymenophore makes it possible to think of *Gastroboletus*, a case of gasteromycetation?

VI. XANTHOCONIUM Sing.

Mycologia 36: 361. 1944.

A combination of yellow to rusty yellow spore print without olive or pink or chestnut tinges, narrow elongated spores, practically always smooth and glabrous to subgelatinous stipe, and unchangeable context characterizes this very homogenous genus. One of its species, has been found in Mexico.

Xanthoconium affine (Peck) Sing. Amer. Midl. Natur. 37: 88. 1947.

Syn.: *Boletus affinis* Peck. Rep. N.Y. St. Mus. 25: 81. 1873.
Suillus affinis Kuntze, Rev. Gen. Pl. 3(2): 535. 1898.
Ceriomyces affinis Murr. Mycologia 1: 149. 1909.
?*Boletus leprosus* Peck, Bull. N.Y. State Mus. 2(8): 135. 1889 (? = var. *maculosus*).

Var. RETICULATUM (A.H. Smith, Persoonia 7: 321. 1973) Wolfe, Can. J. Nat. 6: 2145. Fig. 14-24. 1987.

Pileus from "raw umber" to "Saccardo's umber," Ridgway, then "buckthorn brown" to "tawny olive" or "yellow ocher" to "raw sienna," "buckthorn brown," "honey yellow," "Dresden brown," "raw sienna," "bay," eventually often becoming "Isabella color," subtomentose-subglabrous to glabrous, non-viscid, becoming somewhat rimulose or rimose in age, otherwise smooth or somewhat roughened, pulvinate, then convex with applanate or somewhat depressed center, eventually often irregular, 40-110 mm, broad, rarely broader.

Hymenophore almost whitish when young, then "Isabella color" with or without some "cream buff," especially near the pores, becoming "honey yellow" when mature, unchanging, adnate or more frequently depressed around the stipe; tubes 7-15 mm long; pores concolorous, unchanging or somewhat deeper yellowish or brownish when touched, round or subangular, small (1.4 to a mm) to eventually medium-sized in very large specimens (but still relatively small); spore print between "raw sienna" and "antique brown," or between "Sudan brown" and "Argus brown."

Stipe white, later pallid or pale sordid and often partially "wood brown" or nearer the color of the hymenophore, remaining white or whitish at the base in most individuals, glabrous or subvelutinous, often distinctly reticulated (longitudinal ridges often anastomosing), versiform, either tapering upwards and downwards (ventricose), or tapering from the apex downwards, or thickened as well upwards as downwards (thinnest in the middle), or subequal to equal, solid, 30-40 × 8-25 mm; mycelium inconspicuous, white.

Context white, unchanging or eventually pale ochraceous; odor slight or distinct of *Macrolepiota procera* or "farinaceous", taste mild.

Spores 10.5-17(22) × (2.8)3-5(5.2) μm, with Q = 2.6-4.5, average 4.1, slightly more often fusoid (and then with suprahilar depression) than rod-shaped-cylindrical (like *X. stramineum*), also some oblong, smooth, thin-walled or almost so, slightly less intensely golden or hyaline than in *X. stramineum,* mostly rather a pale golden-yellow (KOH) and weakly pseudoamyloid, with the apex often subacute, not truncate and without a germ pore.

Hymenium: Basidia 24-27 × 7-10 μm, (2)4-spored. Cystidia optically empty, mostly ampullaceous, e. gr. 40 × 9 μm; pseudocystidia internally guttulate and subhyaline, in Melzer ± pseudoamyloid, subventricose, obtuse or mucronate, 36-55 × 9-15 μm, more numerous than cystidia especially at pore level.

Hyphae inamyloid without clamp connections; hymenophoral trama bilateral of the *Boletus*-type. Oleiferous hyphae 4-12 μm.

Covering layers: Epicutis a trichodermial palisade consisting of chains of spherocyst-like and ellipsoid cells 20-40(70) × 18-25 μm with thin- or firm-walls, these often colored bister by intraparietal pigment, the terminal cells often subclavate or attenuated above, without visible contents and inamyloid, in KOH not incrusted. Stipe surface with dermatocystidia which are like the terminal cells of the epicutis chains.

Chemical color reactions: KOH and NH_4OH on context negative or slightly brownish. Other reactions not made on Mexican material. Reactions on Nova Scotian material see Grund & Harrison 1976, p. 83.

In ectomycorrhizal woods under various trees on the ground, in Mexico under *Quercus* and *Pinus*.

Material studied: MEXICO: Nuevo León: Mun. Santiago, El Manzano, 29 VIII 1979, García 41, (F, ITCV).

This variety differs from the type by showing an (often incomplete) reticulation and by having larger, not as Smith stated, smaller, spores than the type variety. This is shown by the type studies by Wolfe (1987). The incrustation on cystidia and epicutis elements is soluble in KOH and visible only in Melzer medium.

The extension of the reticulation of the stipe is probably variable. We suspect that the description of the spores by Smith & Thiers (1971) was taken from specimens that actually belonged to var. *reticulatum* and that the variety exists also in the south of the United States.

The material collected by Castillo & al. (1979) and described by García & Castillo (1981) is described with smooth stipe but large spores. We now rather believe this to be var. *reticulatum* than var. *affine* since the reticulation is often very faint in the former, perhaps at times absent.

Diagnoses latinae

(in alphabetical order)

Austroboletus neotropicalis Sing. García & Goméz spec. nov. Pileo olivaceo-flavo, tomentosulo, 17-35 mm lato; poris roseolis, tactu castanescentibus; stipite alveolato, 60-84 × 5-12 mm; mycelio basali albo; carne castanescente, sicca alba, in vivis amarissima. Sporis (12)13-15 × (5.5)6.6-8 μm ornamentatis verrucis vel cylindris 1.2 μm altis, sed ad basin et apicem brevissimis vel absentibus, inamyloideis. Cystidiis ampullaceis; pseudocystidiis et cheilocystidiis haud differentiatis. Epicute hispida, non gelatinosa. Sub Quercu oleoide in silvis planiteis. Typus: Goméz 18708 (F).

Fistulinella alfaroae Sing. & Gómez spec. nov. PIleo fusco, levi, sicco, 34 mm lato; tubis olivaceis, poris concoloribus, usque ad 1 mm latis; sporis in cumulo purpureo-rosaceis; stipite rubello-brunneo vel fulvo, levi et glabro, solido, 82 × 4-4.5 mm; mycelio basali candido; carne pilei alba vel albida, immutabili vel lenissime sordescente, sed in cavitatibus larvis excavatis rubescente; in stipite superficiei concolori, miti (haud amara), inodora. Sporis 12-16(18.5) × 4-5.5(6.2) μm, fusoideis, levibus, subinamyloideis; cystidiis ampullaceis, contentu destitutis, inamyloideis. Epicute cutiformi, ex cellulis elongatis efformeta; hypodermio trichodermiali. In silva mixta frondosa montana. Monotropae associatus, prope Quercum. Typus Singer, Goméz & Alfaro, Singer B 14621 (F).

Fistulinella wolfeana Sing. & García spec. nov. Pileo roseo, viscido; poris exiguis, tubis immutabilibus; stipite e punctulis brunneis subreticulato, subbulboso, solido; carne (flavido-) alba, immutabili. Sporis 9-13.5(14.3-17) × 4.5-5.5(6) μm; pseudocystidiis numerosis. Ad terram in silvis mixtis (pineis) sed apparenter sub Quercubus. Typus: García 658 (F).

Porphyrellus zaragozae Sing. & García, nov. spec. Pileo griseolofusco vel griseo, 50-85 lato; poris (0.5)1-1.5(2) mm latis; stipite reticulato; carne generatim caerulescente tactu; sporis 13-18(22.5) × 4.3-6(7.5) µm; pseudocystidiis monomorphicis, sed nonnullis pseudoamyloideis, aliis inamyloideis. In silva pineo-quercina, Mexico. Typus: García 2294 (F).

Tylopilus brachypus spec. nov. *T. hondurensi* Sing. & al. peraffinis species, sed differt dermatocystidiis plerumque clavatis neque umquam fusoideis nec ampullaceis et stipite basin versus tantum fortiter attenuato, brevi. Typus in F.

Tylopilus corneri Sing. spec., nov. Pileo castaneo-areolato; poris sat parvis, leniter elongatis radialiter ad marginem pilei; stipite griseolo, levi sed ad apicem ruditer sed subtiliter costulato-subreticulato per 6-7 mm, 52 × 35 mm cc.; carne albida, rubescente fractu. Sporis 9.5-15 × 4-7 µm, levibus et cheilocystidiis sine contentu et inamyloides. Epicute palisadica trichodermiali e cellulis elongatis efformata, intus porphyreis sed inamyloideis. Sub Quercubus in silvis montanis, Costa Rica; typus: Gómez & Alfaro 20582 (F).

Tylopilus costaricensis Sing. & Gómez spec. nov. Pileo purpureo vel violaceo-brunneo; stipite ad ipsam apicem tantum vel haud vere reticulato, obclavato vel obclavato-appendiculato. Carne haud rosacente fractu. Sub Quercubus in silvis montanis. Typus in F. — *T. ammiratii* affinis.

Tylopilus gomezii Sing. spec. nov. Pileo umbrino vel dilute umbrino. Poris brunneolis fractu. Stipite coriicolori vel castaneo-brunneo, non vel ad apicem tantum reticulato. Forma et magnitudine *T. lividobrunneo* simillimus est sed differt praecipue carne rosascente vel purpurascente. Sub Quercubus in silvis montanis, typus in F.

Tylopilus jalapensis Sing. & García spec. nov. Pileo roseologriseo, in juvenilibus saepe magis purpureo-violaceo-tincto. Stipite albo vel albido, levi et glabro. Carne immutabili amaro. Sporis 6.3-10 × 3.3-4.5 µm; cheilocystidiis et pseudocystidiis moderatim numerosis; trichodermio in parte superiore haud palisadico; stipite obtecto strato hymeniali vel epitheliali quod e cellulis sphaericis vel late ventricosis efformatum est. Sub Quercubus in Mexico. Typus in F conservatus.

Tylopilus mitissimus Sing. & Gómez spec. nov. Pileo strato trichodermiali palisadico obtecto. Carne pilei alba, immutabili. Sub Quercubus zonae montanae in Costa Rica. Typus in F.

Tylopilus montoyae Sing. & García spec. nov. Pileo rosello-brunneo, tomentoso, 60 mm lato. Tubulis grisello-roseis, brunnescentibus, poris concoloribus, parvis. Stipite pileo subconcolori, apice subtiliter reticulato, ceterum pruinato, ± 110 × 9-10 mm. Carne alba vel albida, fractu nigricante. — Sporis 10.8-12.8 × 4.6-5.8 μm. Cystidiis contentu rubello-brunneo. Epicute in juvenilibus trichodermio palisadico formata, cellula terminali frequenter cystidis hymenialibus simili. Typus Montoya-Bello 699 (F).

Tylopilus subcellulosus Sing. & García spec. nov. Pileo brunneo, 75-100 mm lato. Stipite subconcolori vel subtiliter violaceo tincto, flocculis retiformiter dispositis minutis ornato; carne immutabili, amara. Sporis plerumque 10.5-11.5 × ± 3.5 μm; cystidiis et pseudocystidiis praesentibus; trichodermio pilei 100-300 μm diam., in parte superiore saepe palisadico. Sub Quercubus in Mexico, typus in F.

Tylopilus subniger Sing., García & Gómez spec. nov. Pileo primum subnigro, demum violaceo-brunneo; stipite atro-reticulato per tertiam apicalem vel ex toto; carne miti ut minime partim fractu rosascente vel roseolo-grisascente. Sporis plerumque 11-14 × 4-5.5 μm; pseudocystidiis numerosis, cystidiis paucis. Epicute trichodermiali, haud palisadica. Sub Quercubus in silva montana. Typus in F.

Tylopilus vinaceogriseus Sing., García & Gómez spec. nov. *T. ferrugineo* var. *vinaceogrisea* Snell, Dick & Hesler nec non *T. costaricensi* peraffinis species, sed ab hoc stipite manifeste venoso-reticulato, ventricoso et carne manifeste rosascente differt; ab illo differentia vix patet cum varietas paullum descripta est. A *T. ferrugineo* var. *ferrugineo* differt colore pilei magis violacino vel obscuriore et cystidiis dermatocystidiisque pilei paulum angustioribus. Habitat: Sub Quercubus in silvis montanis. Typus in F.

Tylopilus williamsii Sing. & García spec. nov. Sporis 8-9 × 4 μm a *T. subcelluloso* differt. Sub Quercubus in Mexico, typus in F.

Acknowledgements

We acknowledge the John D. and Catharine T. MacArthur Foundation (grant number 875101) which was awarded to G.M. Mueller (F) for the partial support of our field work in Costa Rica.

J. García wishes to thank CONACYT for support of his work on Mexican Boletineae. He also thanks the curators of XAL, ENCB and FCME, and particularly Sr. Alfredo González, ENCB, who kindly made available part of this thesis material.

Literature cited in part III

CORNER, E.J.H. (1972): Boletus in Malasia. - Government Printing Office Singapore.

GARCÍA, J. & J. CASTILLO (1981): Las especies de Boletaceos y Gomfidiaceos conocidas en Nuevo León. - Bol. Soc. Mex. Mic. **15:** 121-197. 1981.

GARCÍA, J., G. GAONA, J. CASTILLO & G. GUZMÁN (1988): Nuevos registros de Boletaceos en México. - New records of Boletaceae in México. - Rev. Mex. Mic. **2:** 343-366.

HALLING, R.E. (1983): Boletes described by Charles C. Frost. - Mycologia **75:** 70-92.

HALLING, R.E. (1989): A synopsis of Colombian boletes. - Mycotaxon **34:** 93-113.

MOORE, R. & T. GRAND (1970): Application of scanning electron microscopy to basidiomycete taxonomy. - Proc. 3rd SEM Symp.

PEGLER, D.N. (1983): Agaric flora of the Lesser Antilles. - London.

SINGER, R. (1945): The Boletineae of Florida 197-141. (Reprint J. Cramer 1977).

SINGER, R. (1966): Die Röhrlinge I. - Klinghardt, Bad Heilbrunn.

SINGER, R. (1986): Agaricales in modern taxonomy. 4. ed., Koeltz Scientific Books.

SINGER, R. (1988): La fitogeografiá de las Boletineas (Basidiomycetes, Agaricales) based on Mexican species. - Rev. Mex. Mic. **4:** 267-274.

SINGER, R., I. ARAUJO & M.H. IVORY (1983): The ectotrophically mycorrhizal fungi of the neotropical lowlands, especially Central Amazonia. - Nova Hedw., Beihefte **77:** 1-339. J. Cramer, Vaduz.

SINGER, R. & R.E. MACHOL (1977): Are Secretan's fungus names valid? - Taxon **26:** 251-255.

SMITH, A.H. & H.D. THIERS (1968): Notes on boletes I. - Mycologia **60:** 943-954.

SMITH, A.H. & H.D. THIERS (1971): The boletes of Michigan. - Ann Arbor.

SNELL, W.H. & E.A. DICK (1970): The Boleti. - J. Cramer, Lehre.

WELDEN, A. & G. GUZMÁN (1978): Lista preliminar de los hongos, liquenes y Myxomicetos de las regionens de Uxpanapa, Coatzacoalcos, los Tuxtlas, Papaloapan y Xalapa (parte de los estados de Verycruz y Oaxaca. - Bol. Soc. Mex. 2: 59-102.

WOLFE Jr., C.B. (1979): *Austroboletus* and *Tylopilus* subgenus *Porphyrellus* with emphasis on North American taxa. - J. Cramer, Bibliotheca Mycologica, **69.**

WOLFE Jr., C.B. (1981): Type studies in *Tylopilus* I. Taxa described by Charles H. Peck. - Sydowia **34:** 199-213.

WOLFE Jr., C.B. & PETERSEN (1978): The type of *Boletus fumosipes.* - Mycologia **70:** 676-679.

Supplement to Part I

A tropical variety of *Paxillus atrotomentosus* which we consider an independent species has never been found in Mexico or Central America until it was collected by Ovrebo in Costa Rica and determined as identical with *Paxillus atrotomentosus* var. *bambusinus* by Singer. It is here cited as *Paxillus bambusinus* (Baker & Dale) comb. nov. although it is certainly not restricted to Bambusinae and probably not ectomycorrhizal so that the name is somewhat embarassing. We add a key to the species of sect. *Atrotomentosi* occurring in Central- and South America:

1. Cystidia absent or indistinct, context without orange and/or violet stains..... 2
1. Cystidia present, distinct; context with orange, violet stains. *P. amazonicus* Sing.
    2. Pileus surface guttate. Brazil.......................... *P. guttatus* Sing.
    2. Pileus not guttate. Trinidad to Central America... *P. bambusinus* (B. & D.)

Paxillus bambusinus (Baker & Dale) Sing. comb. nov.

Syn.: *Paxillus atrotomentosus* var. *bambusinus* Baker & Dale, Myc. Pa. C.M.I. 32. 1951.

Pileus dark buff when young, becoming dull gold in age with purplish brown, scattered, minute fibrillose clumps which are more distantly spaced toward the margin, otherwise appearing glabrous, dry to moist, spathulate or petaloid in outline when young, becoming orbicular or occasionally subdimidiate, pale to broadly convex, eventually plane to slightly concave, the margin often uplifted in age 30-110 mm broad. Spore deposit on surface of pileus dull brown (not obtained on paper).

Lamellae dull gold or honey colored (closest to 5C5*) often in places with a faint yellowish cast, especially near the margin of the pileus, entire, subdistant (through-lamellae up to 3 mm apart), lamellulae numerous, but tiers not distinct, 3.5-5 mm broad, subdecurrent.

*Color chart indications from A. Kornerup & J.H. Wanscher, London 1978.

Stipe buff but below the buff ring of the apex only with a buff ground color, above this dark purplish brown (8-9BF 5) tomentose, with the fibrillose tufts more distantly spaced at apex, solid, eccentric, occasionally lateral, equal, 20-50 × 4-10 mm.

Spores 4-4.5 × 2.5-3 more rarely only 3.5 μm long or up to 5 μm long and up to 4.5 μm broad, ellipsoid, more rarely ovoid, the adaxial side often less convex, more applanate than the dorsal one, smooth, pale yellowish brown, without germ pore and without suprahylar depression with a thin, inamyloid wall and a slightly elongated oil drop.

Hymenium consisting of basidia which are 17-18-(27) × (4.5)-5-(6) μm, 4-spored, few 2-spored ones present then 4-4.5 μm broad.

Hyphae with clamp connection. Hymenophoral trama bilateral, the mediostratum regular, 3.5-9,3 μm broad, brownish hyaline with hyphae more interwoven to almost interlaced, but many rising at an almost right angle from the mediostratum, not visibly gelatinized.

Covering layers: Epicutis of pileus consisting of a loosely entangled cutis whose hyphae are 3-6 μm diam. with scattered cylindric to clavate end cells, 27-40 × 6-12 μm, all cells smooth, thin-walled, hyaline to pale yellow. Covering of stipe consisting of loosely interwoven, 3.5-6 μm diam., thin-walled, pale yellow hyphae. Hyphae below the uppermost layer 4-15 μm broad.

Gregarious and often cespitose-imbricate on fallen dicotyledonous log.

Material studied: COSTA RICA: Heredia. Biological Station "La Selva", 29 VII 1989, Ovrebo 281 (F).

We believe that this taxon should be accepted on the species level. *P. atrotomentosus* has somewhat larger spores, more purplish brown tomentum on the stipe, and different habitat, viz. on dead wood of Angiosperms rather than on wood of conifers. Although some pilei of *P. bambusinus* in Costa Rica reach 110 mm diam., the average size of this species is smaller, and the pilei realtively thinner. The distribution of the two species seems to be mutually exclusive. *P. bambusinus* has been observed in Trinidad (Baker & Dale, l.c., Dennis) and is said by Pegler (1983) to occur in Honduras. Some of the original notes by Dr. Ovrebo were used in the description, but this covers only the material from the collection 281.

Index to part III (and supplement)

Sections, subgenera and genera

Austroboletus 4, 53
Austroboletus (sg., sect.) 53

Fistulinella 3, 73
Fistulinella (sect.) 73, 80

Gastroboletus 34
Graciles (sect.) 53, 56

Ixechinus 73

Leccinum 3
Leucogyroporus 8

Mucilopilus 73

Paxillus 94
Phaeoporus 60
Porphyrellus 60
Porphyrellus (sg.) 60
Porphyrosporus (sg.) 60
Pseudotylopili (sect.) 79
Pulveroboletus 4, 5

Rhodobolites 8
Rhodoporus 8

Scrobiculatae (sect.) 73, 74, 84
Scrobiculati (sect.) 74

Tristes (sect.) 53
Tylopilus 3, 4, 8

Veloporphyrellus 3

Wolfeanae 73, 82

Xanthoconium 4, 85

Species and varieties

affine 85
affine (var.) 87
affinis 85
alboater 9, **20,** 50
alfaroae 74, **77,** 88
amarus (var.) 47
amazonicus 94
ammiratii 42
amylosporus 60
areolatus 25
atrobrunneus 68
atrotomentosus 94, 95

badiceps 19
balloui 8, 12
bambusinus 94
belizensis (var.) 74, 76
brachypus 12, **47,** 89
brunneirubens 68
brunneus 80

conica 74
conica (var.) 74, 76
corneri **50,** 89
costaricensis 11, **36,** 89
cyaneocinctus (ssp.) 64
cyaneotinctus 61, 64

dictyotus 53

ferrugineus 11, 28, **42,** 46
fuligineus 68
fumosipes 66, 67

glabripes (var.) 67
gomezii 10, **30,** 89
gracilis 56, **58**
griseopurpureus 25, 47
griseus 70
guanacastensis 10, **27**
guttatus 94
guzmaniana **73,** 83

heterospermus 56
hondurensis 10, **46,** 49

indecisus 12, **44**

jalapensis 9, **16,** 35, 89
jamaicensis 73

leprosus 85
lividobrunneus 10, **29,** 32

major 81
mexicana 80
mexicanus 80
minor (Fist.) 81
minor (Tylop.) 25
mitissimus 9, **17,** 89
montanus 22
montoyae 9, **19,** 89

nebulosus 68
neofelleus 25
neotropicalis **53,** 55, 88
nicaraguensis 9, **24**
niger 70
nigerrinus 40
nigrellus 20

obscurus 9, 22
olivaceobrunneus 70

pachycephalus 51
pacificus 61
pantoleucus 5
phaeocephalus 51
plumbeoviolaceus 9, **14,** 35
porphyrosporus 61, 67, **71**
pseudodecorus 37
pseudoscaber 68

reticulata (var.) 77
reticulatus (var.) 85

sanctaerosae 10, **25**
snellii 64, 68
sordidus 64, 67, 68, 72
subflavida 54
subcellulosus 11, **32,** 90
subniger 11, 27, **31,** 90
subpunctipes 28
subunicolor 51
subvirens 55

tabacinus 47

umbrosus 60, **62**

vinaceogriseus 11, **40,** 90
vinaceogriseus (ssp.) 42
vinosobrunneus 35
viscidus 79

williamsii 11, **33,** 35, 90
wolfeana **82,** 88

zaragozae 60, **69,** 88.

Plates

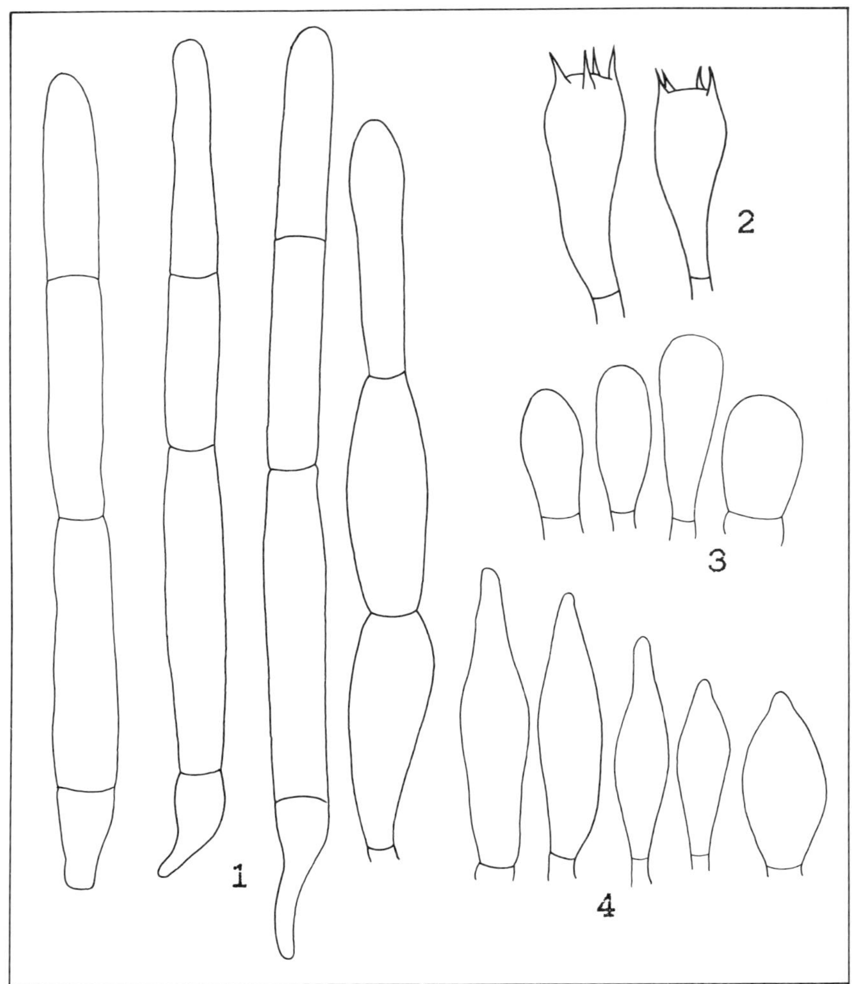

Plate 1: *Veloporphyrellus pantoleucus*. Mexican collection. 1. Elements of the covering of the stipe. — 2. Dermatobasidia (stipe. — 3. Dermatobasidioles (stipe). — 4. Dermatocystidia (stipe). All ×660.

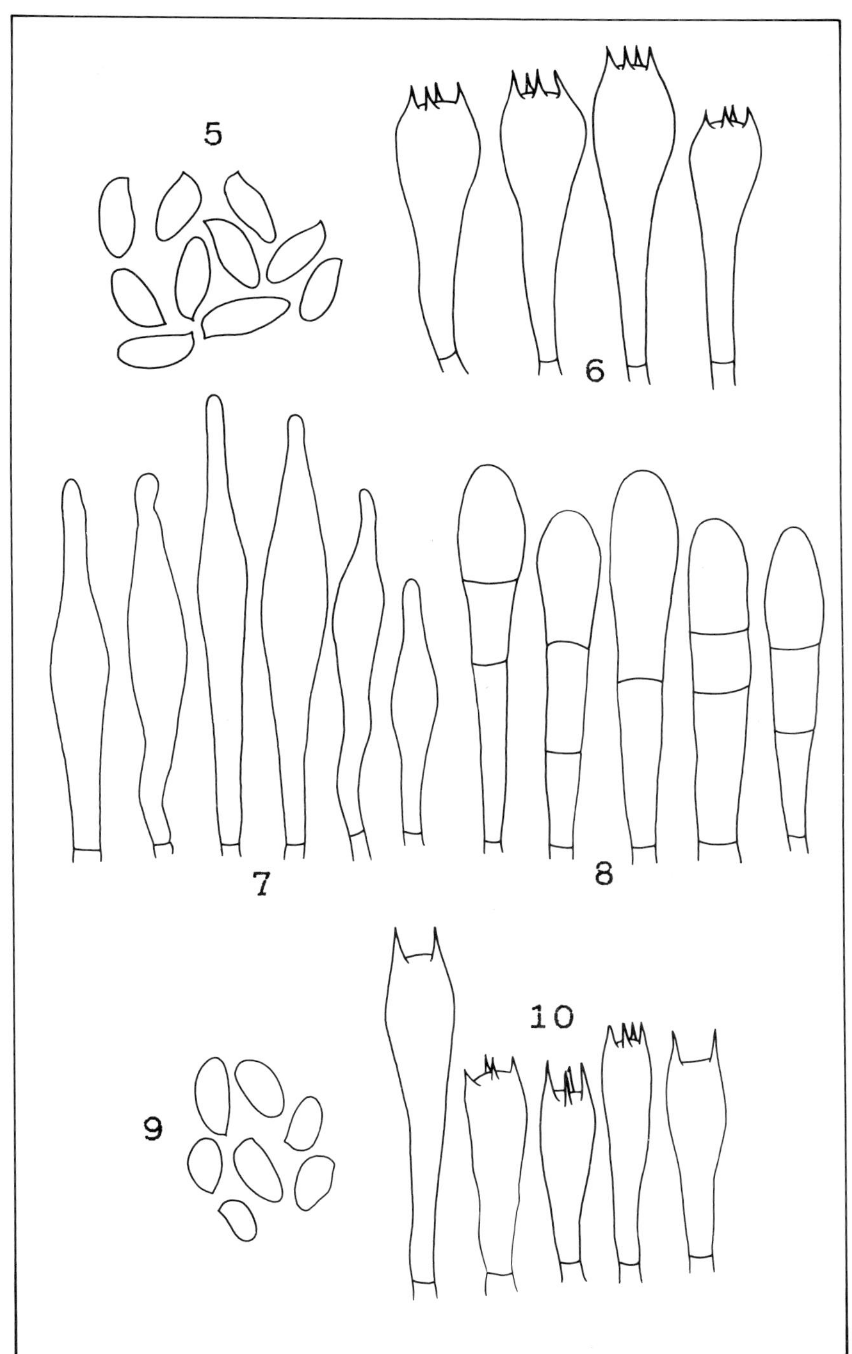

Plate 2: *Veloporphyrellus pantoleucus*. Mexican collection. 5. Spores. — 6. Basidia (hymenium). — 7. Cystidia. — 8. Other elements of the lamellar edge. All ×660. *Tylopilus balloui*. 9. Spores ×990. — 10. Basidia ×660.

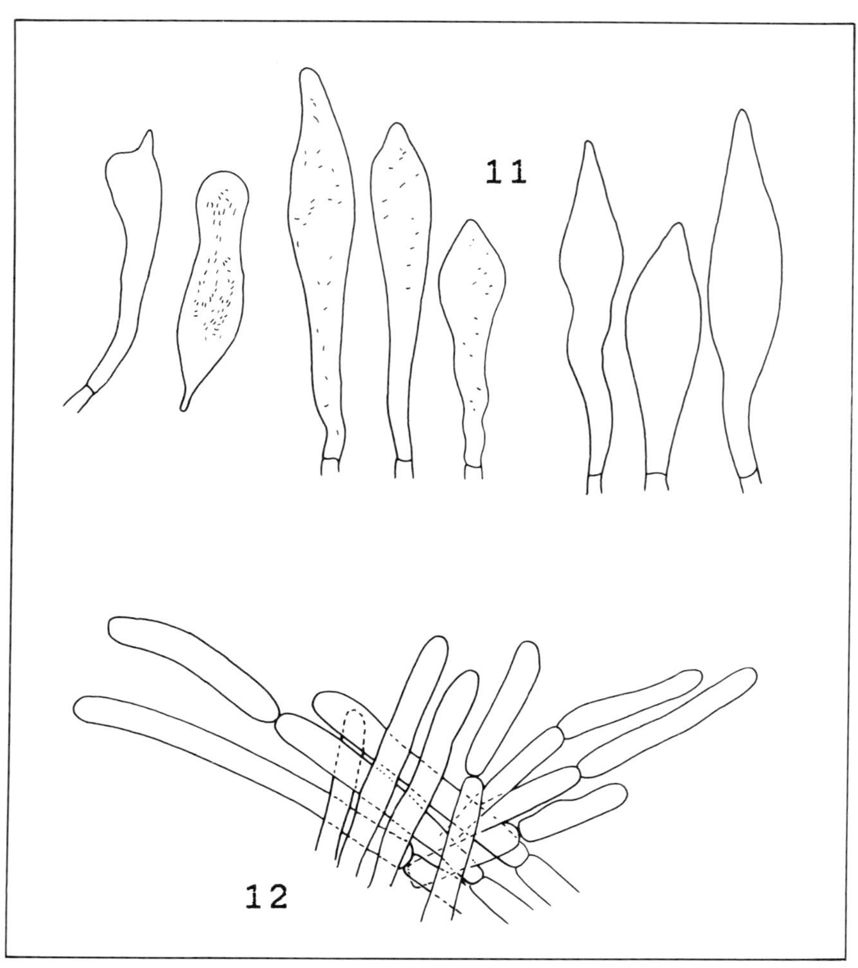

Plate 3: *Tylopilus balloui*. 11. Cystidia ×990. — 12. Epicutis ×660.

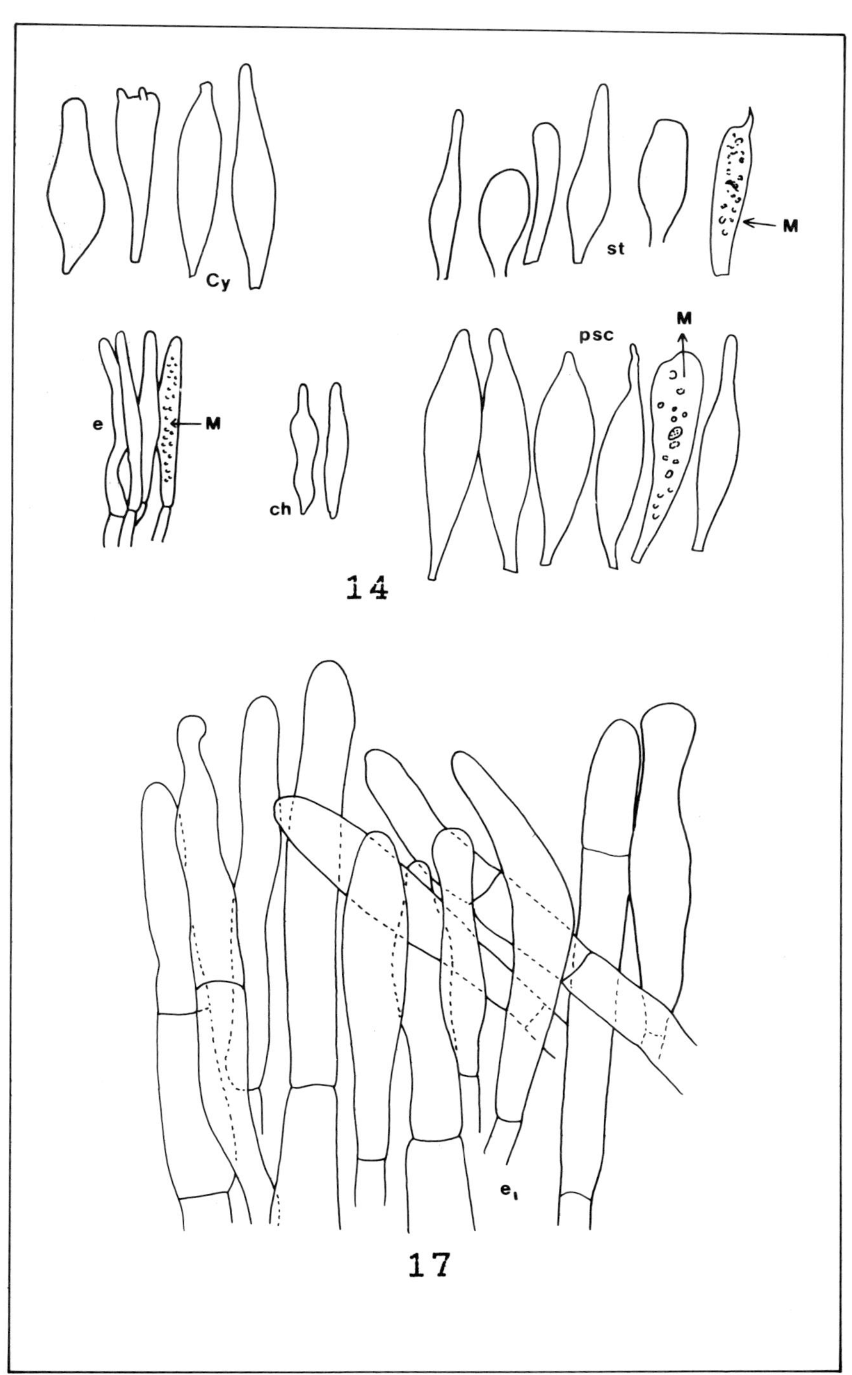

Plate 4: *Tylopilus mitissimus*. 14. cy = cystidia; st = elements of stipe covering; e = elements of epicutis; ch = cheilocystidia; psc = pseudocystidia (hymenial); M = cystidiform cells with guttulae. In psc: ×860. Others ×440.
Tylopilus alboater. 17. e = epicutis ×660.

Plate 5: *Tylopilus mitissimus*. Fig. 15. Spores ×320 and carpophore ×2/3.
Tylopilus alboater. Fig. 16. Carpophore ×2/3.
Porphyrellus cyanotinctus. Fig. 62. Cheilocystidia ×660.

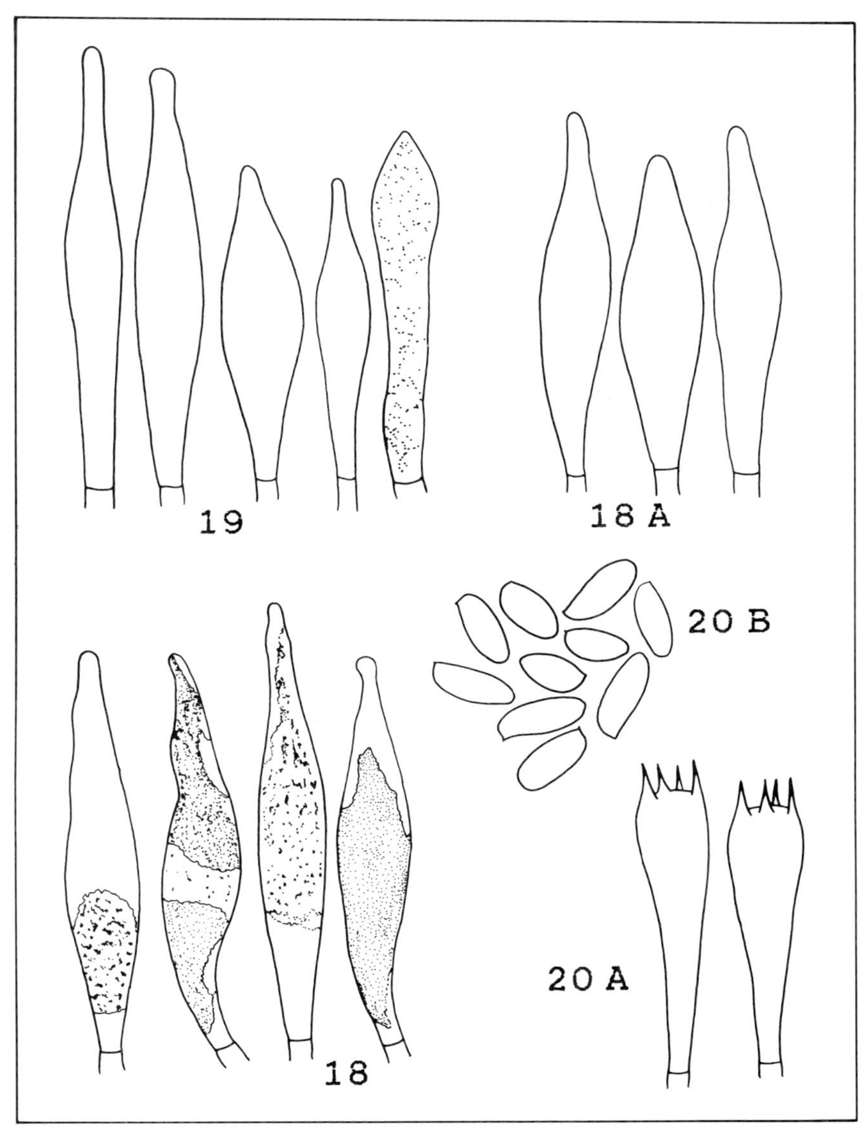

Plate 6: *Tylopilus alboater*. 18. Pseudocystidia. — 18A. Cystidia. — 19. Cheilocystidia. — 20A. Basidia. — 20B. Spores. All ×660.

Plate 7: *Tylopilus obscurus*. 21. Carpophore × 2/3. — 22. Left: Dermatocystidia of epicutis, the one marked M with contents. Right: pseudocystidia and cystidia of hymenium. All × 660.
Tylopilus guanacastensis. 23. Two carpophores (left no. 18725; right no. 21316) × 2/3; spore × 1340 and two pseudocystidia × 670.
Tylopilus lividobruneus. 24. Carpophore × 2/3 and spores × 2680.
Tylopilus lividobruneus, type. 25. above: pseudocystidium, midle: cheilocystidia, below: dermatocystidia. All. × 670.

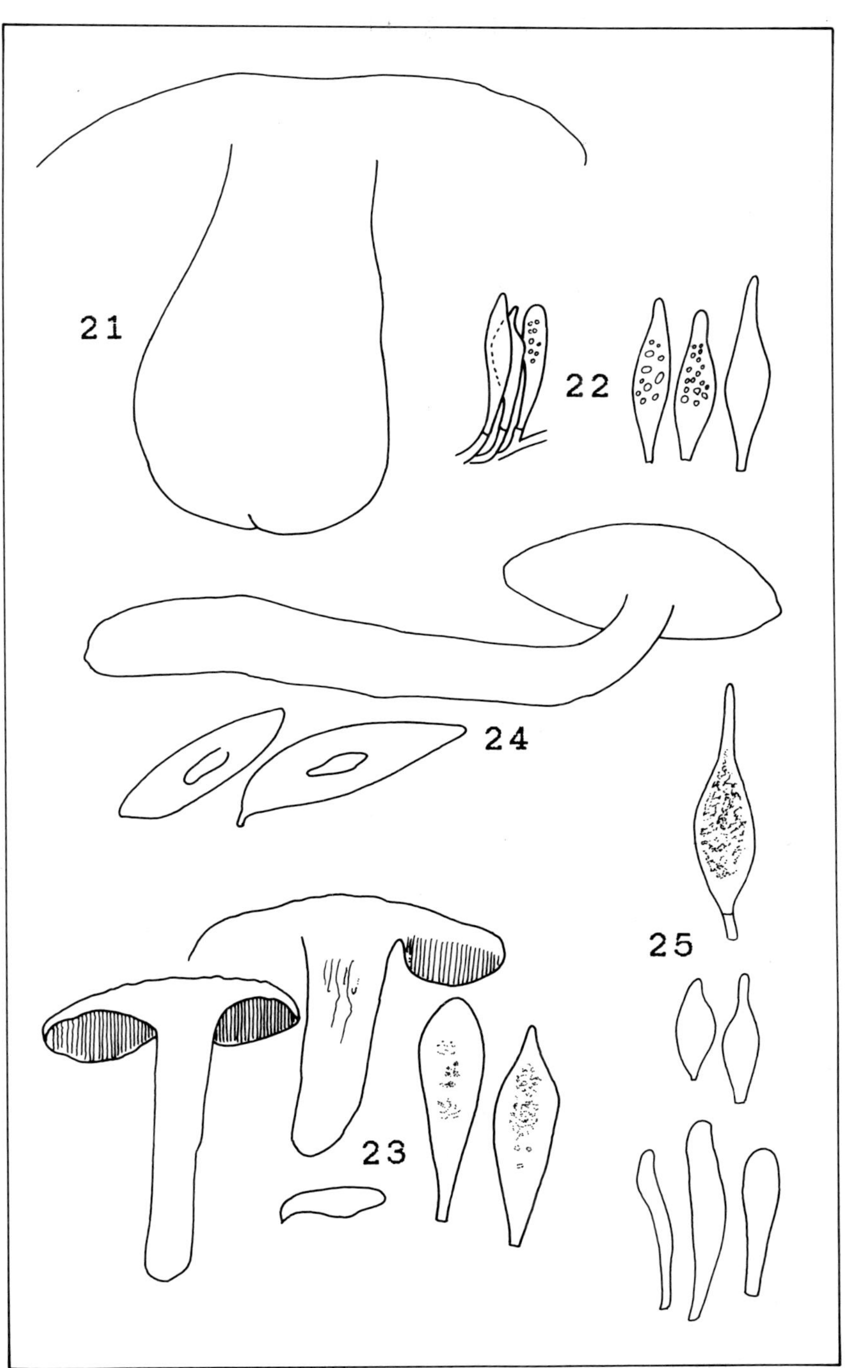
21
22
24
25
23

Plate 8: *Tylopilus gomezii*. 26. Two carpophores (left no. 22018; right no. 21453) × 2/3, (left: from stipe, right from hymenium, all × 660, and spores × 1320. *Tylopilus subcellulosus*. 27. Elements of epicutis, on extreme right one cystidium, all × 660, and spores × 1320.

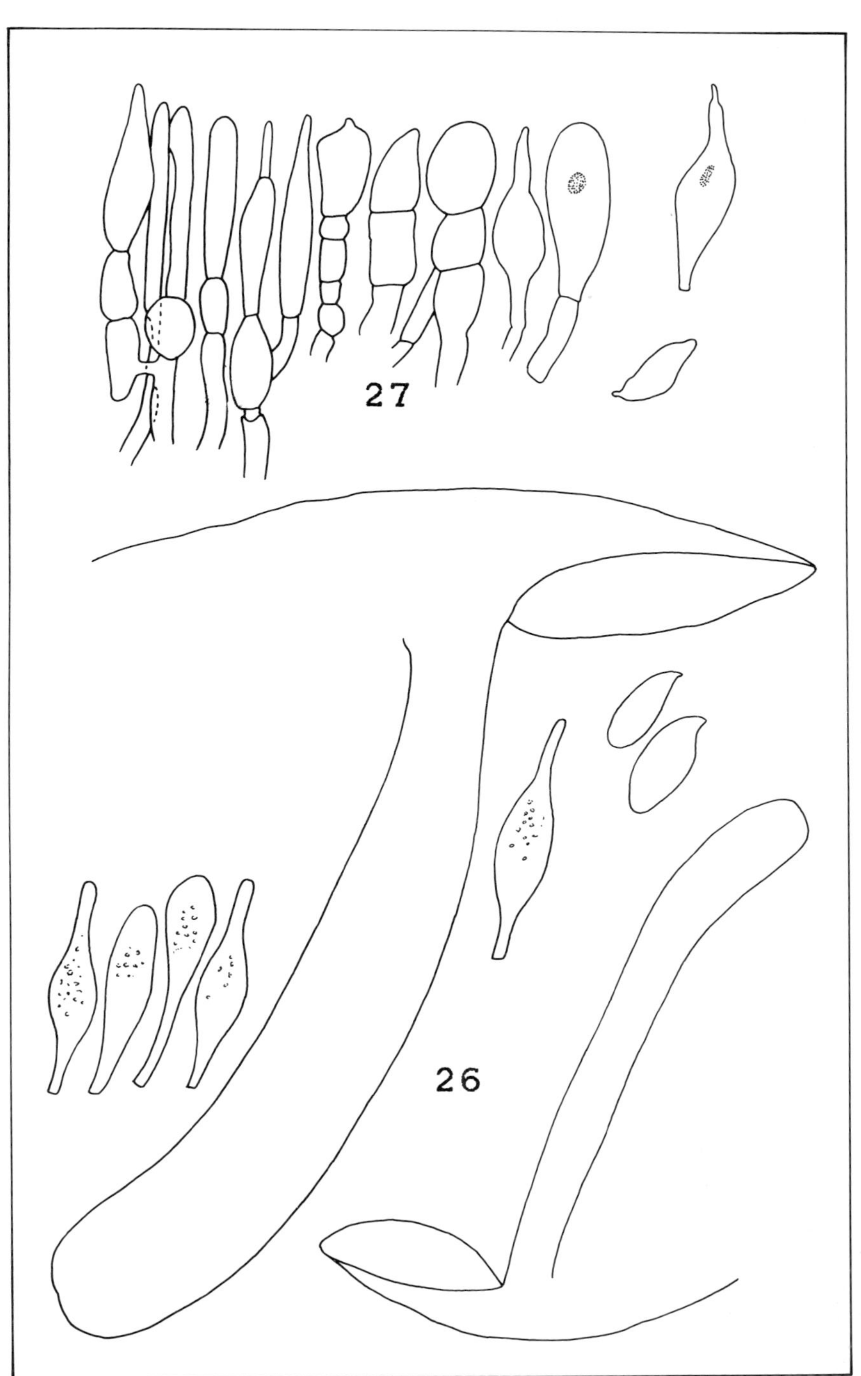
27
26

Plate 9: *Tylopilus costaricensis*. 28. Two carpophores, left no. 20621, right no. 20626, both ×1. Right, from above: three pseudocystidia; hyphae of epicutis (left no. 20626, right no. 20621); cystidia, 20626 (left), and dermatocystidia from stipe (right), all ×660. Spores (left below) ×1320, a giant spore among them.
Tylopilus subniger. 29. Above: Dermatocystidia of pileus and (right) elements of the covering of the stipe, both ×660. Below one carpophore, ×2/3, and cystidia of the hymenium, ×660.
Tylopilus williamsii. 30. From above: Epicutis ×660; spores ×1320, two hymenial pseudocystidia and one cystidium, ×660.

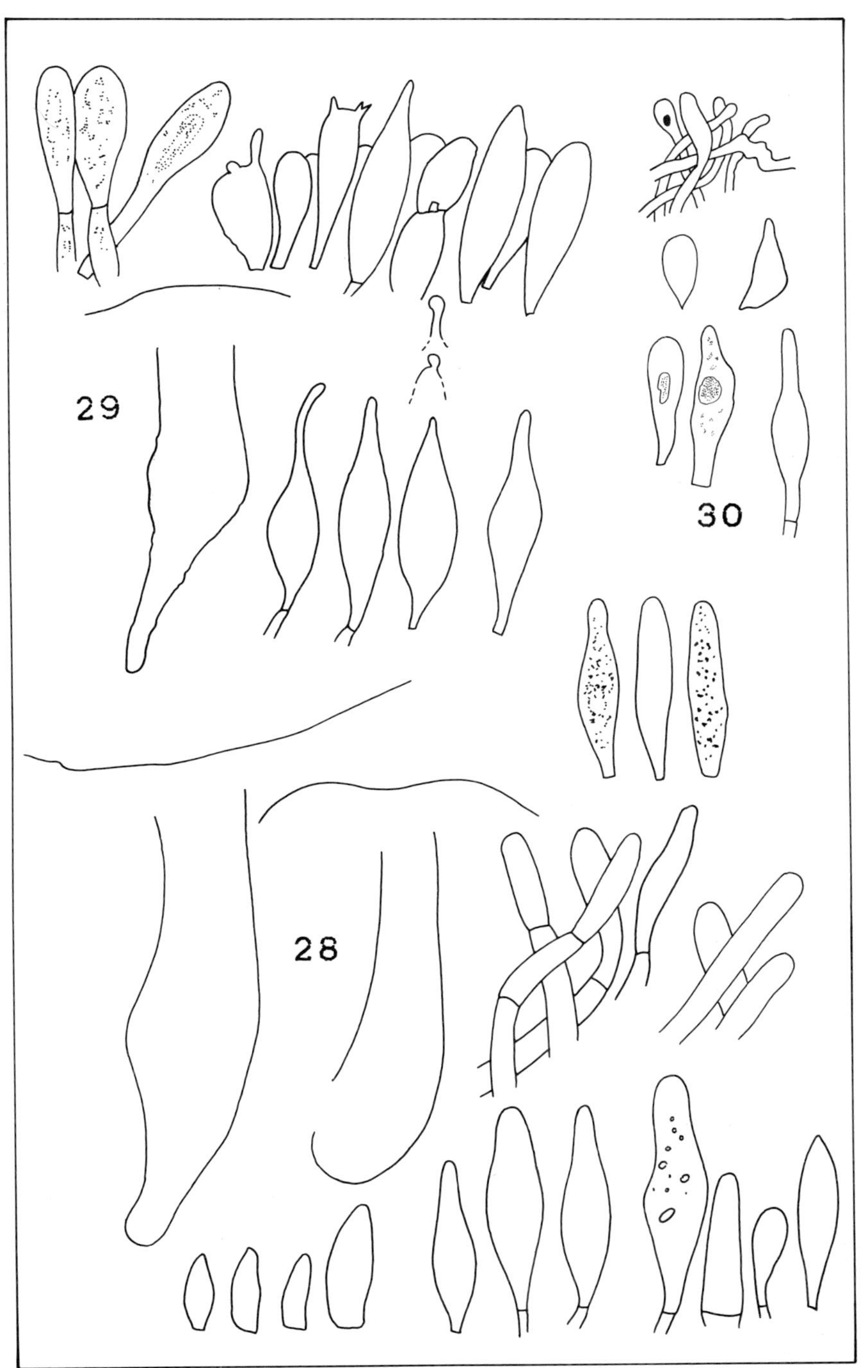
29
30
28

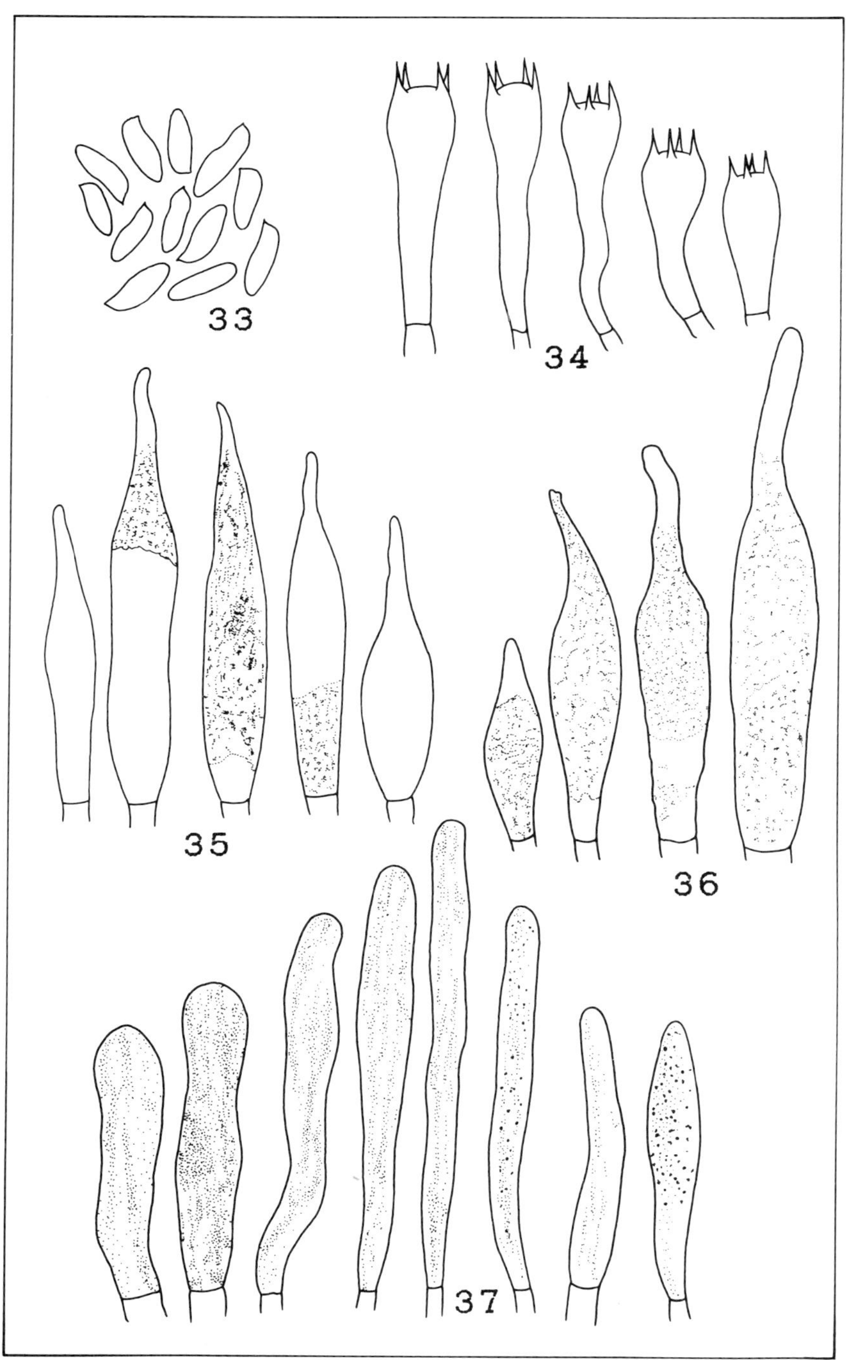

Plate 10: *Tylopilus vinaceogriseus*. 33. Spores. — 34. Basidia. — 35. Cystidia from pores. — 36. From interior of tubes. — 37. Elements of epicutis. All ×660. Many of them pseudocystidial.

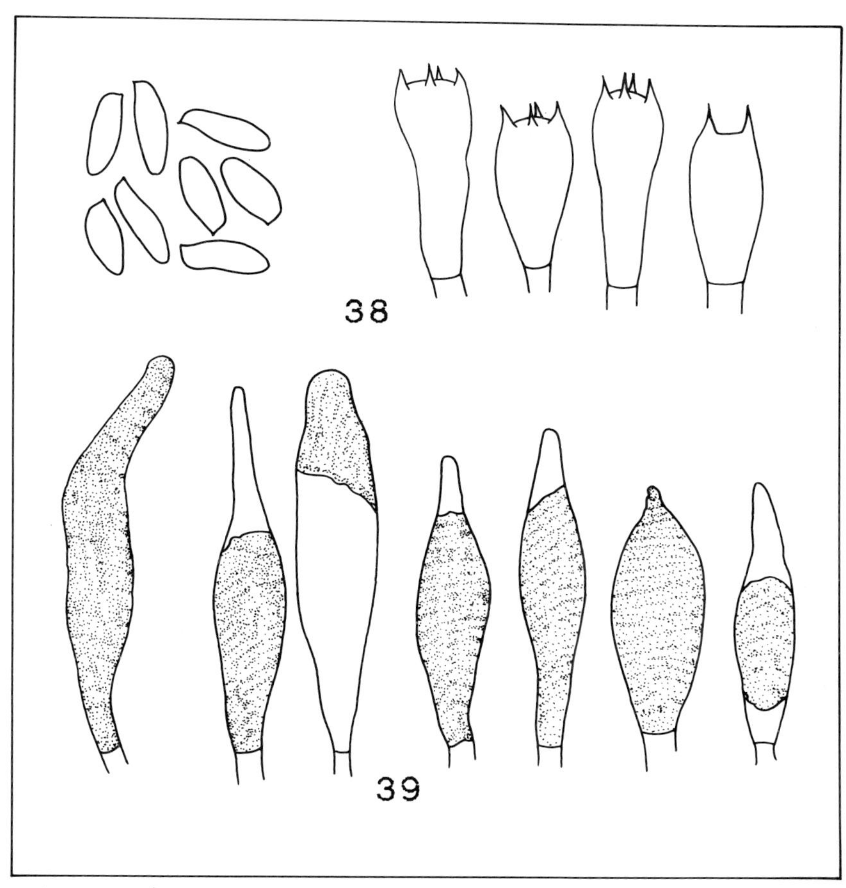

Plate 11: *Tylopilus subniger*. 38. Spore and basidia. — 39. Pseudocystidia. All ×660.

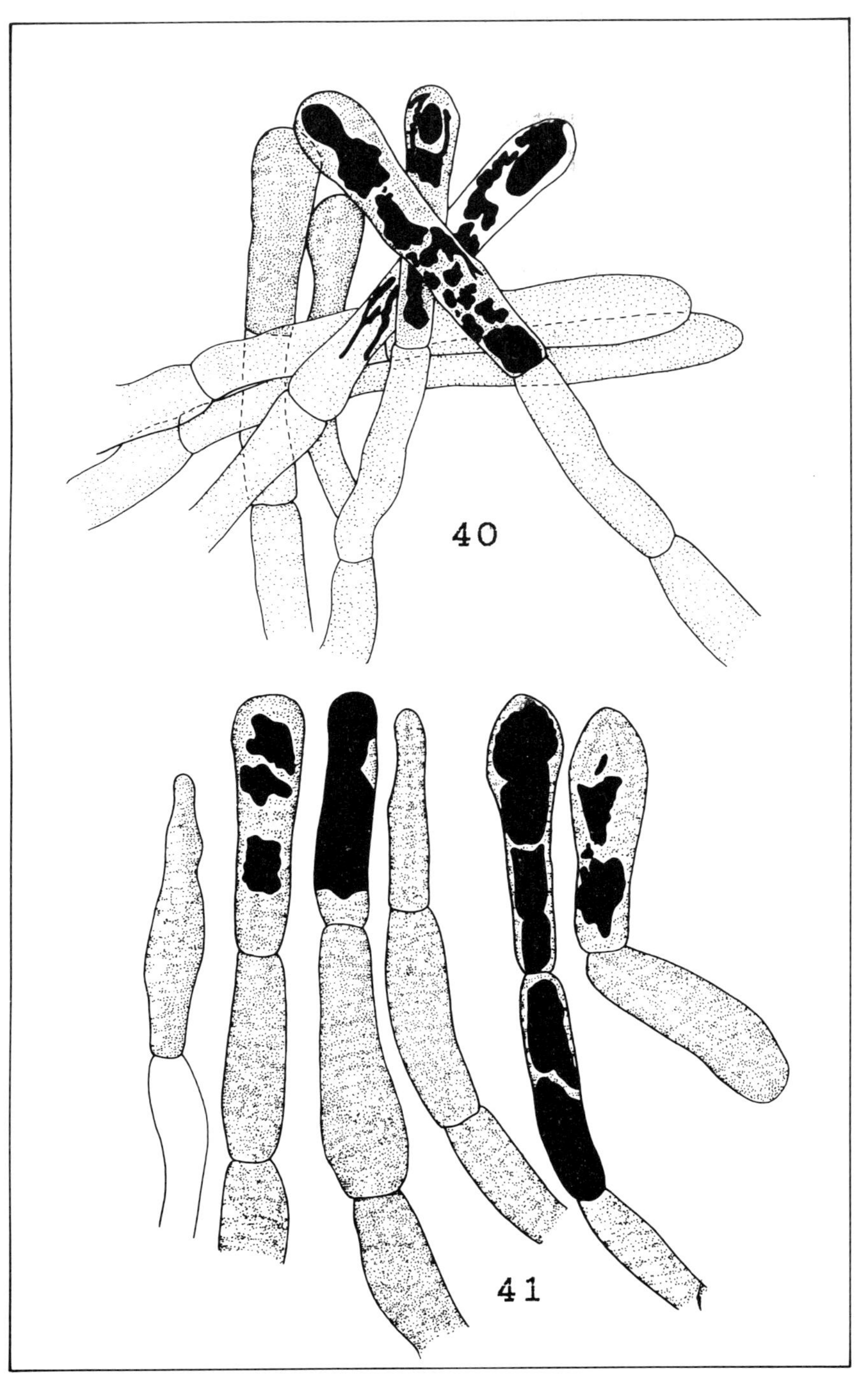

Plate 12: *Tylopilus subniger*. Figs. 40, 41. Elements of epicutis ×660. Both this and preceding plate from Mexican material.

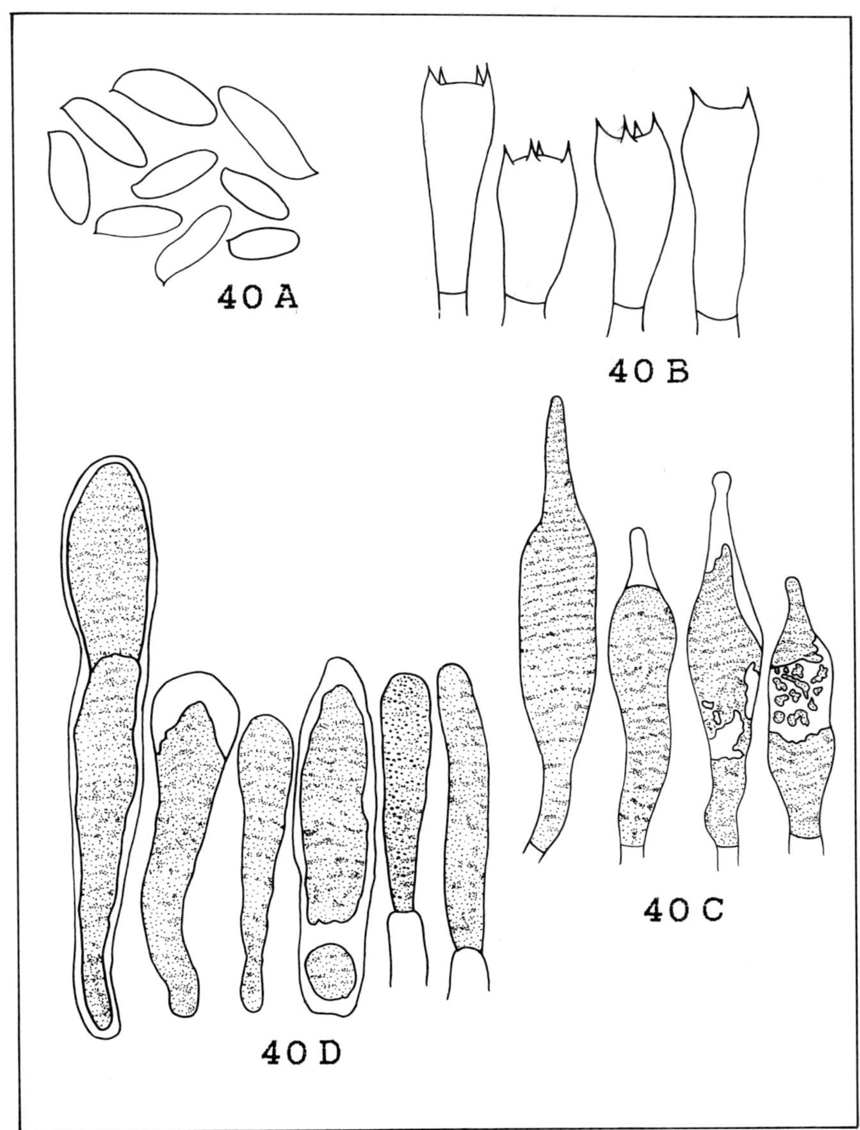

Plate 13: *Tylopilus subniger*. Figs. 40A. Spores ×1400; 40B. Basidia ×1000; 40C. Pseudocystidia ×100; 40D. Terminal cells of the epicutis ×1000. This is from material collected in Costa Rica.

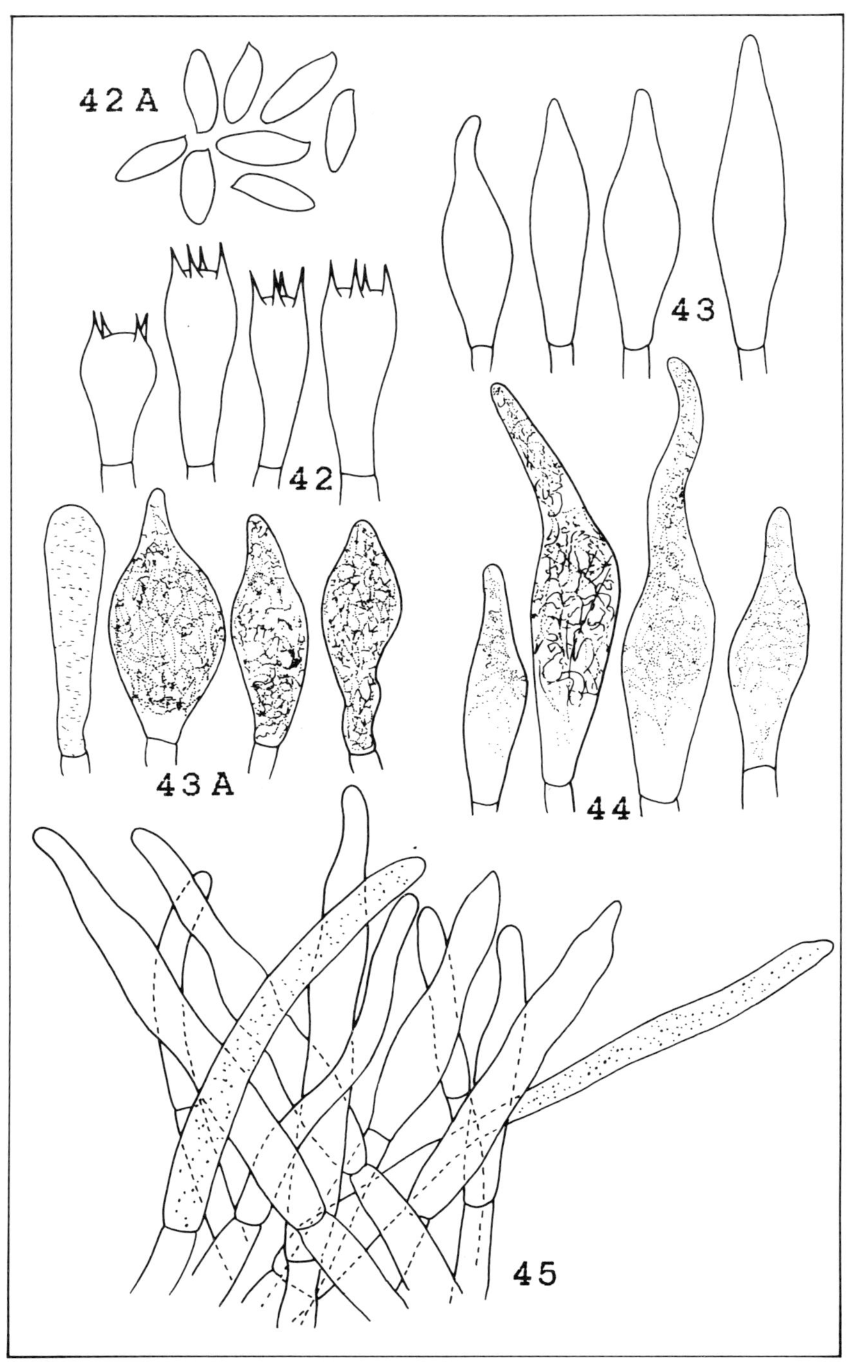

Plate 14: *Tylopilus ferrugineus.* 42. Basidia; 42A. Spores. — 43, 43A. Pseudocystidia, the four at the left as seen in Melzer's reagent. — 44. Some cystidiform cells on the reticulum of the stipe. — 45. Epicutis. All ×660.

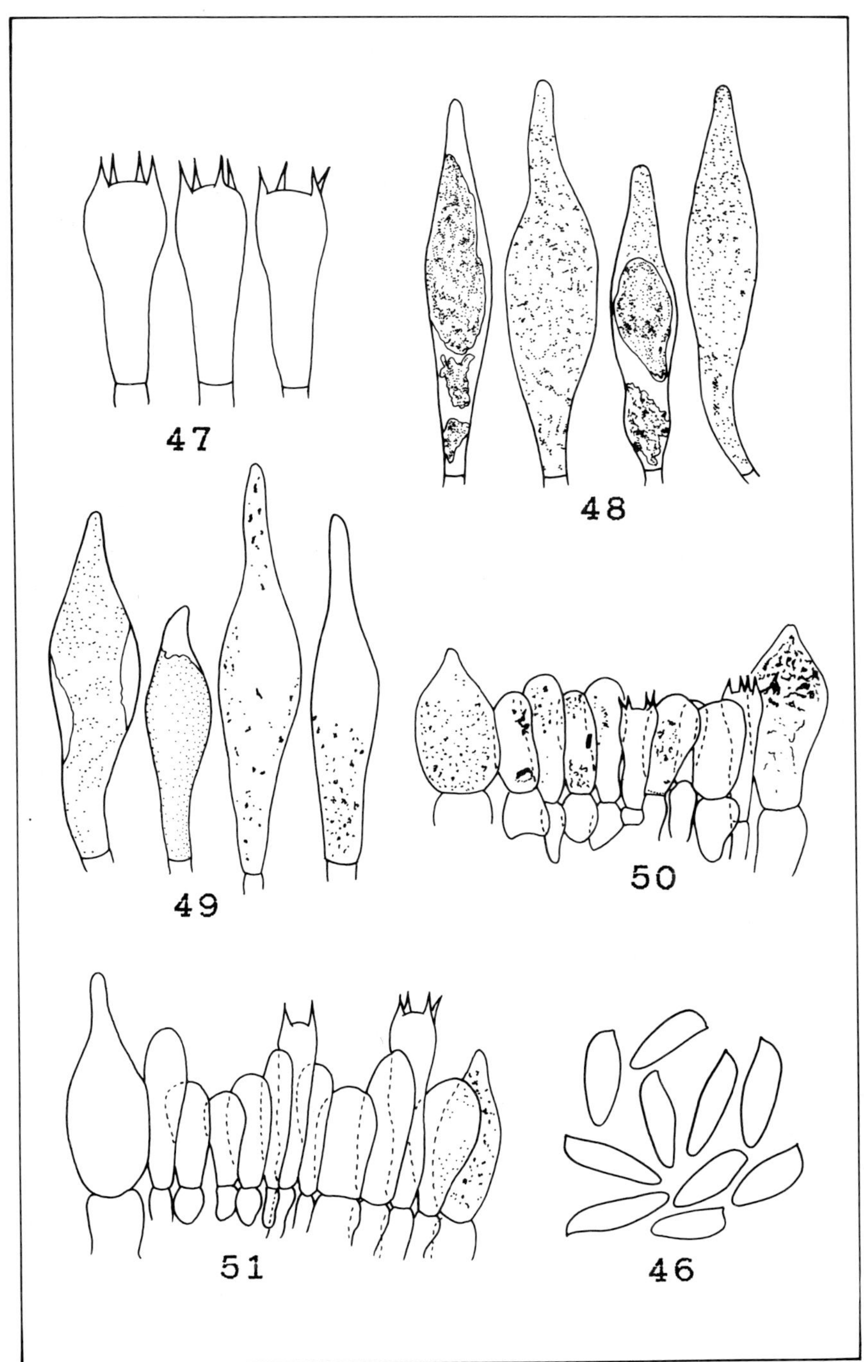

Plate 15: *Tylopilus brachypus*. 46. Spores. — 47. Basidia. — 48, 49. Pseudocystidia. — 50. Surface elements of the stipe. — 51. Surface of the stipe. 46-49 ×660, 50-51 ×440.

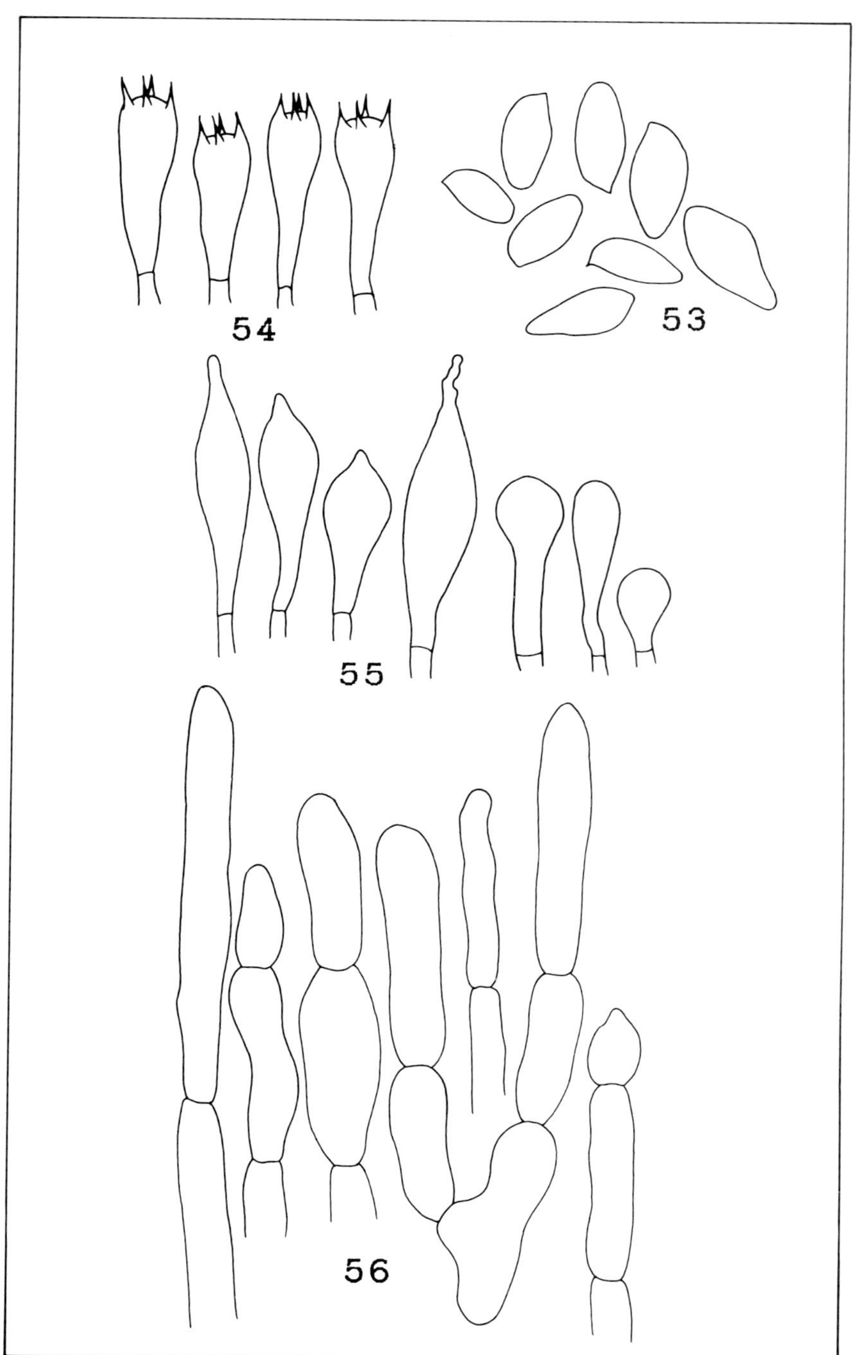

Plate 16: *Porphyrellus pacificus*. 53. Spores ×1000. — 54. Basidia. — 55. Cystidia. — 56. Epicutis. 54-56 ×440.

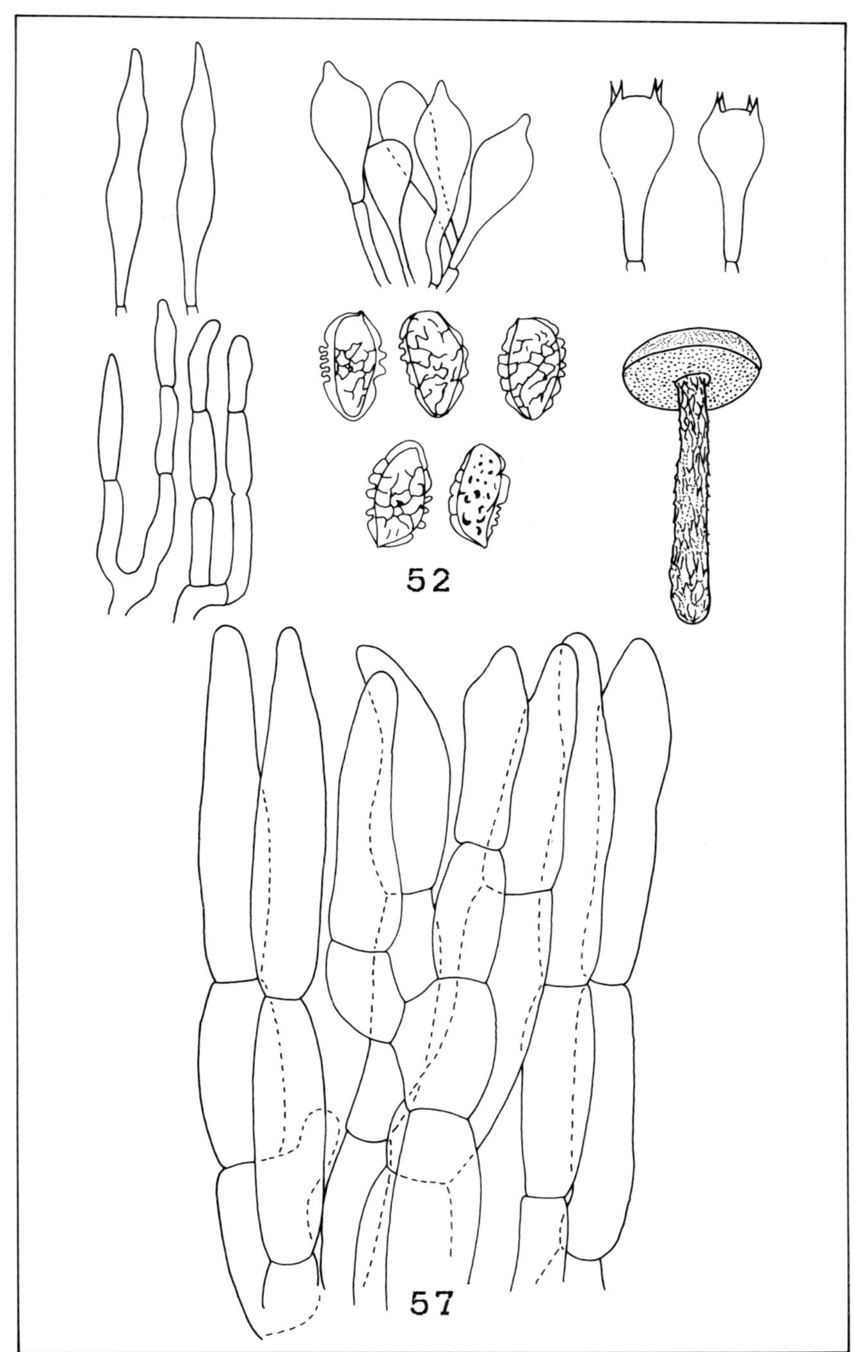

Plate 17: *Austroboletus neotropicalis.* 52. Upper left: Eurocystidia. Center: Elements of the network of the stipe. Right: Basidia. Lower left: Epicutis. Cener: Spores. Right: Carpophore.
Porphyrellus umbrosus. 57. Epicutis ×1000.

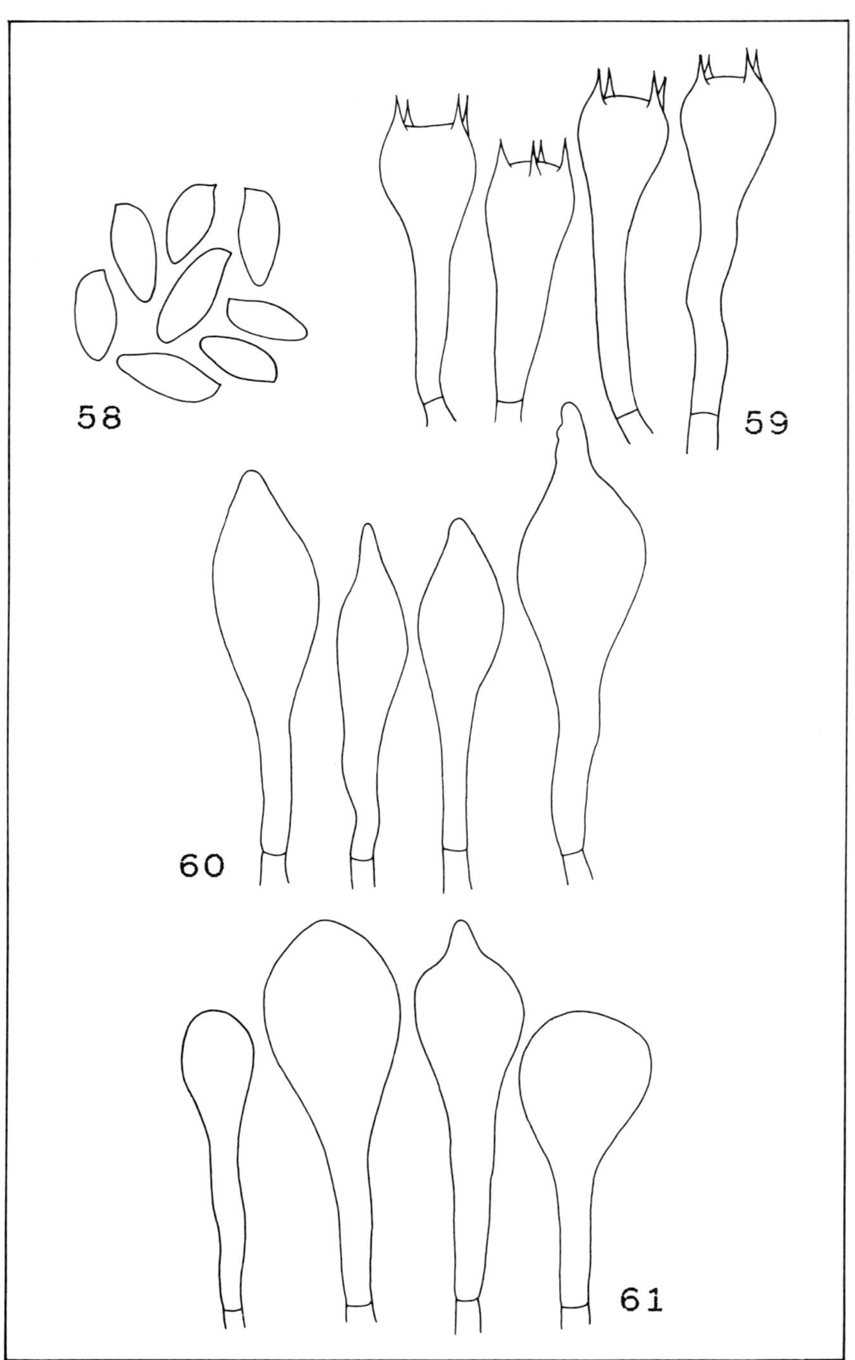

Plate 18: *Porphyrellus umbrosus*. 58. Spores. — 59. Basidia. — 60. Pleurocystidia. — 61. Cheilocystidia. All ×660.

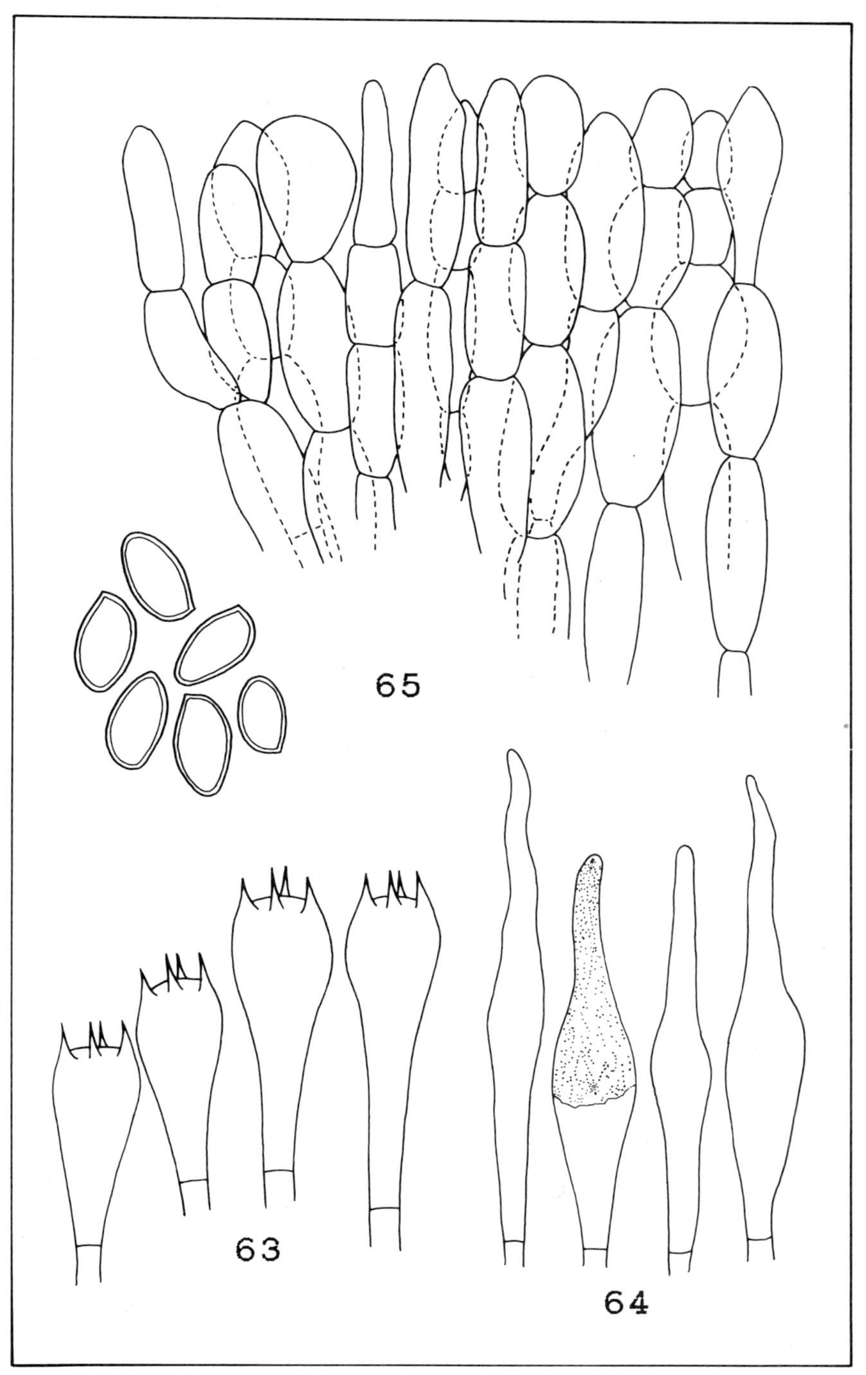

Plate 19: *Porphyrellus cyaneotinctus*. 63. Basidia ×1000. — 64. Cystidia ×1000. — 65. Left, below: Spores × 1000 ; right: Epicutis, the latter ×490.

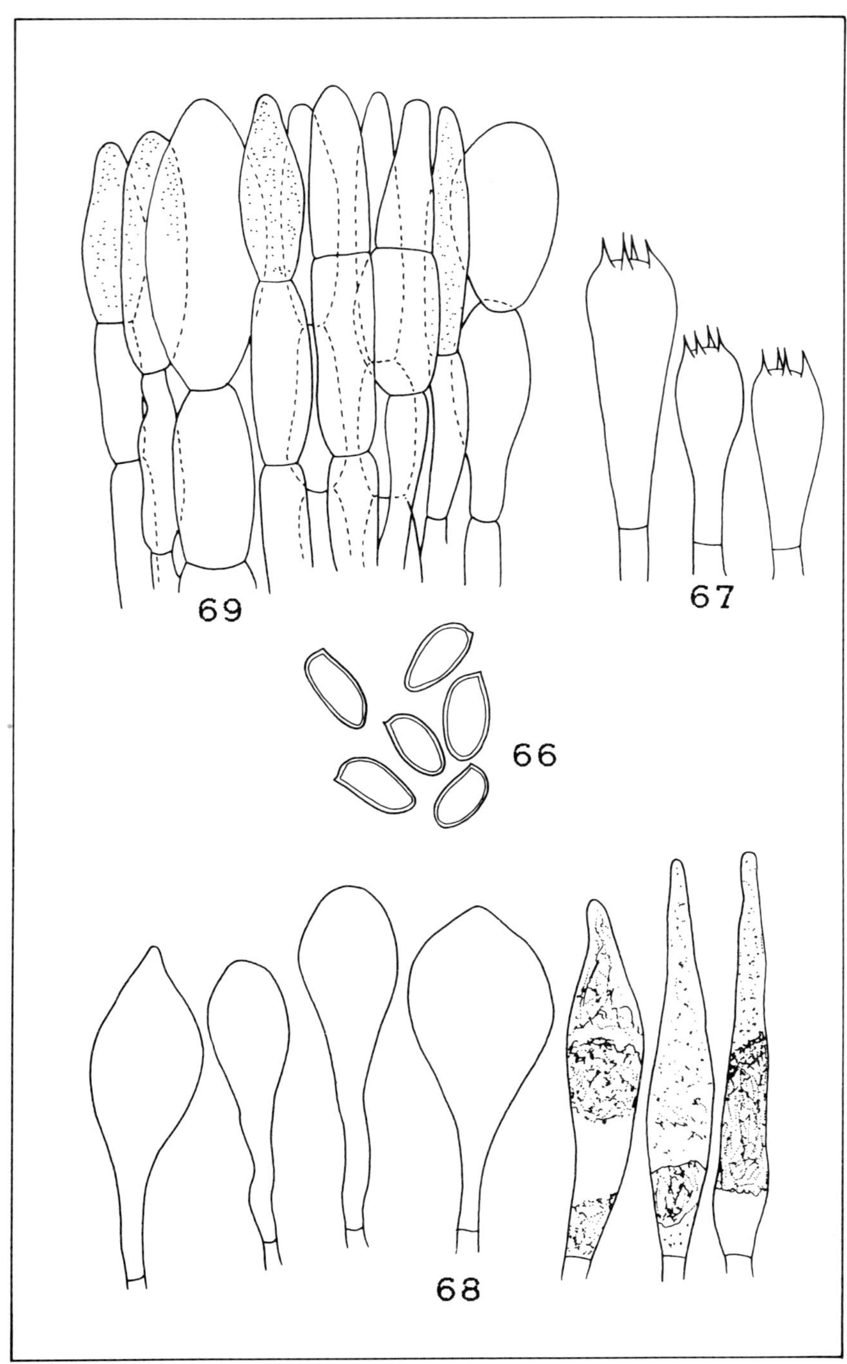

Plate 20: *Porphyrellus cyaneotinctus*. 66. Spores. — 67. Basidia. — 68. Cheilo- and pleuro-(pseudo)cystidia. — 69. Epicutis. All ×660.

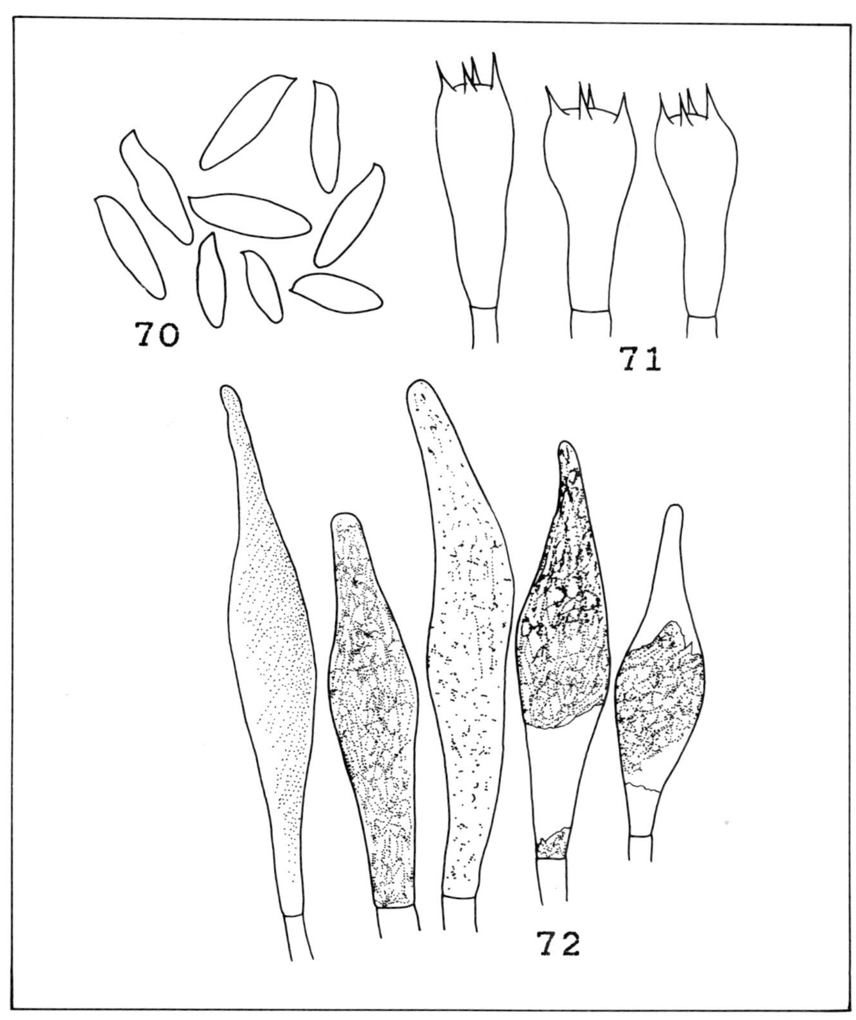

Plate 21: *Porphyrellus zargozae*. 70. Spores. — 71. Basidia. — 72. Pseudocystidia. All ×660.

Plate 22: *Tylopilus jalapensis*. 13. Cheilocystidium (left) and elements of stipe covering both ×660.
Tylopilus vinaceogriseus. 31. Carpophore ×2/3; 4 elements of stipe covering (upper right) and three (isolated) pseudocystidia, upper from pores, all ×660. From Costarican material.
Fistinella alaroae. 76. Carpophore ×2/3; below left: Cystidia; middle: Elements of the surface of the margin of the pileus ×660; right surface of the stipe, ×660.
Fistulinella guzmania. 77. Section of carpophore ×2/3.
Fistulinella wolfeana. 78. Spore ×1330, below from left to right one dermatocystiduum and two dermatopseudocystidia of the stipe, epicutis of the pileus, four hymenial (pseudo-)cystidia, ×660.
Fistulinella mexicana. 79. Spores ×1330.

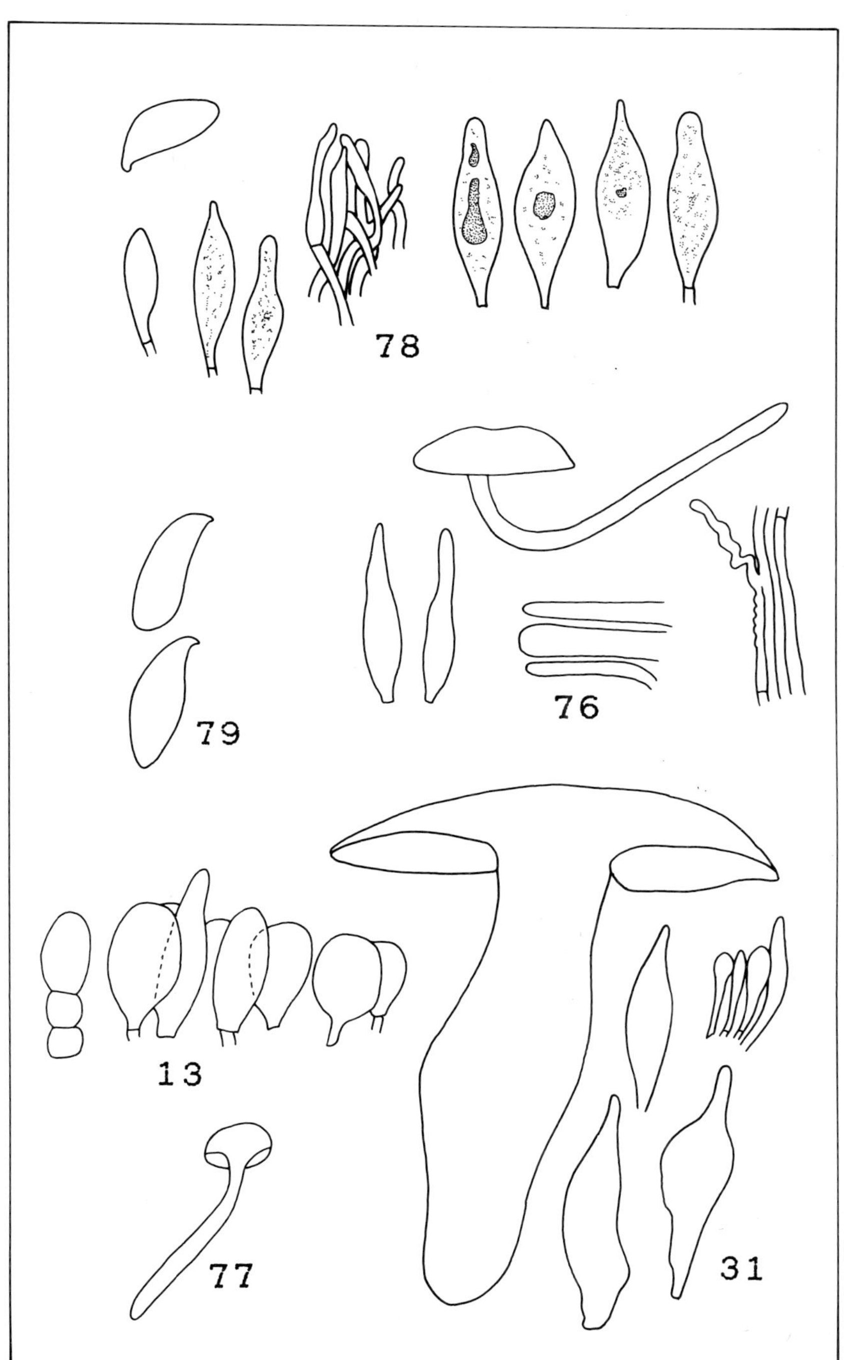
78
79
76
13
77
31

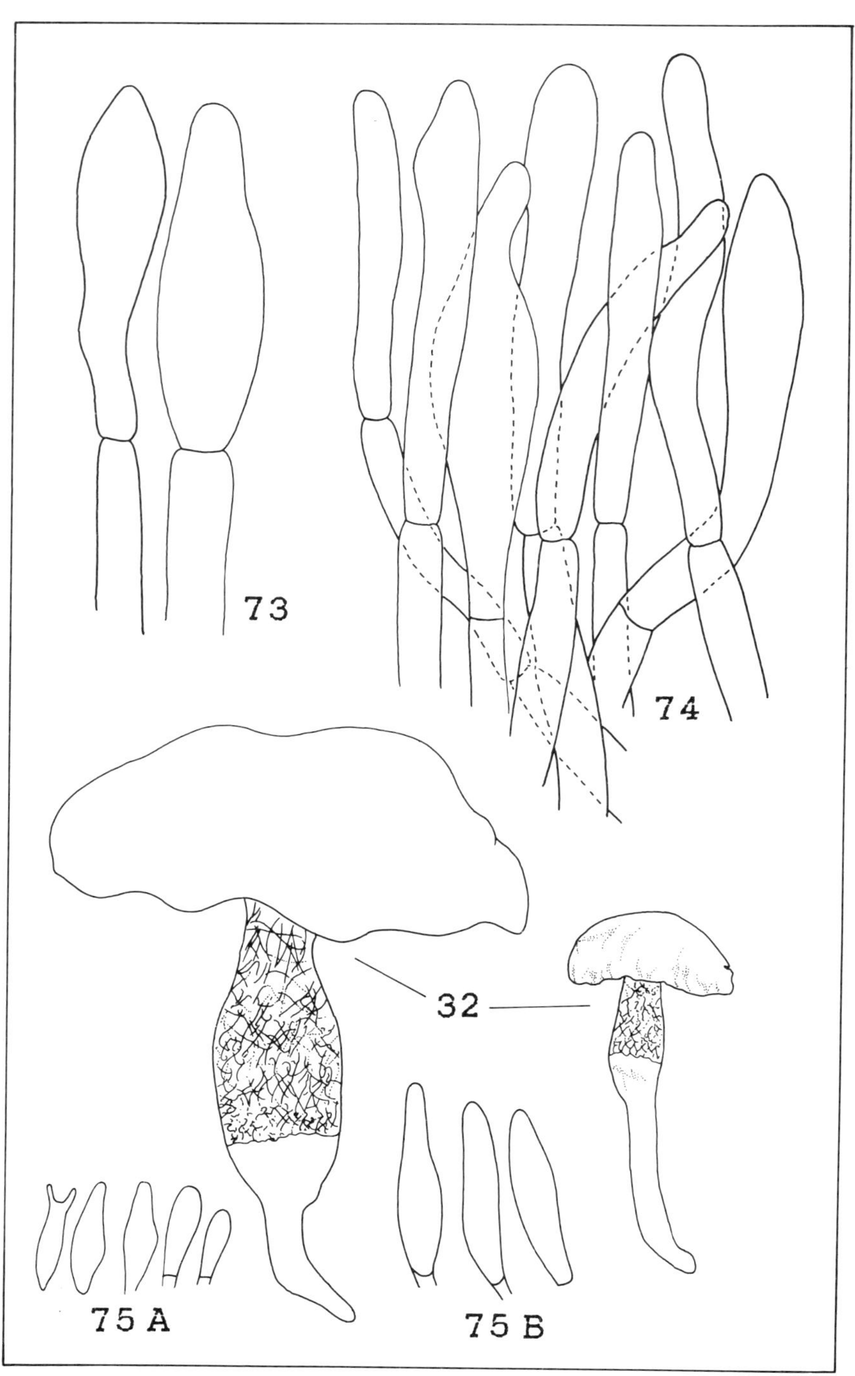

Plate 23: *Porphyrellus zaragozae*. 73. Spores. — 74. Basidia. — 75. Pseudocystidia. All ×1000.
Fistulinella conica var. *belizensis*. 75A. 5 elements of stipe covering; 75B. 3 Cystidia. All ×660.
Tylopilus vinaceogriseus. 32. Dried carpophores ×2/3; Mexican material.

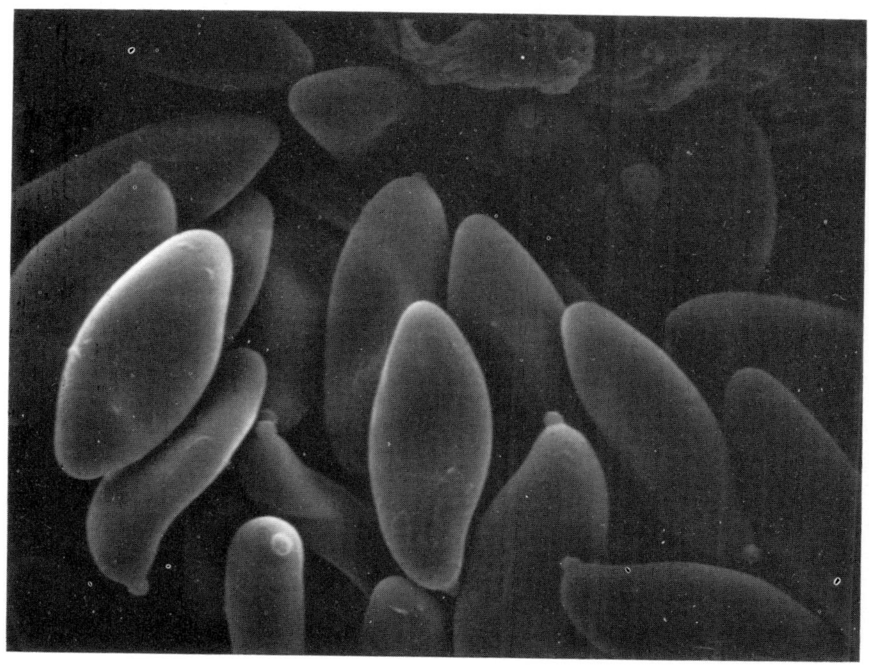

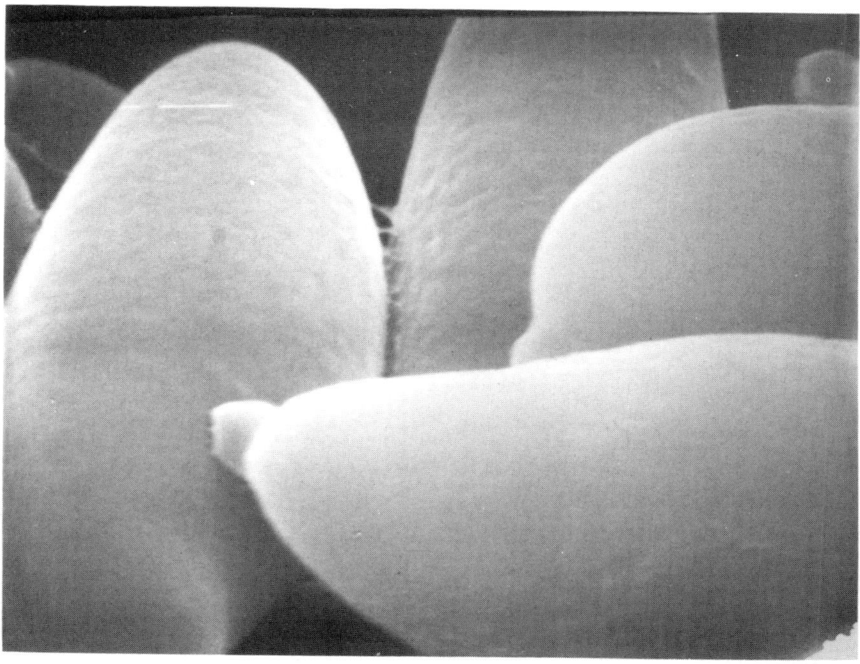

Plate 24: *Fistulinella alfaroae.* Upper SEM ×3300, phot. Wolfe jr., from type; lower, same but 10 000×.